The BMW 2002

a comprehensive guide to the classic sporting saloon

cabriolet · saloon · ti · tii · touring · turbo

James Taylor and Mike Macartney

The BMW 2002
A Comprehensive Guide to the Classic Sporting Saloon Car

Designed by Peter Bevins

Cover Photography by Colin Burnham. Reproduced by kind permission of Classic Car magazine.

Brooklands Books

ISBN: 9781783181964
4W0/2771

Brooklands Books are dedicated to preserving motoring literature for enthusiasts. They have in print over 700 titles which deal with various aspects of motoring.

Brooklands Books Ltd., PO Box 904,
Amersham, Bucks, England. HP6 9JA

E-mail: sales@brooklands-books.com
www.brooklandsbooks.com

Contents

This book is dedicated to all BMW 2002 enthusiasts throughout the world.

Introduction

This is an unusual type of book for Brooklands Books to publish, and we are pleased to have been asked to write it. We both learned a great deal from the process, and we hope our readers will find it meets their needs and interests.

We did of course receive a lot of enthusiastic help in putting this book together. Thanks therefore must go to AC Schnitzer, Alpina Burkhard Bovensiepen GmbH & Co, BMW AG in Germany, BMW (GB), Peter Bevins, Martin Bull, Brendon Lopez, Rob Torres Jr of the BMW Autohaus and the dozens of BMW 2002 enthusiasts whose experiences contributed to the opinions and advice given in the later Chapters of the book. Brooklands Books are also pleased to thank the publishers of the original road test material reproduced in Chapter 6 for their kind co-operation in this project.

James Taylor Mike Macartney

M-CS 868

Chapter One

BMW Before the 2002

BMW - the Bayerische Motoren Werke or Bavarian Motor Works - can trace its car-making history back as far as 1928. But the firm itself dates back to 1916 when the Bayerische Flugzeug Werke (Bavarian Aircraft Works) was founded in Munich to support the Kaiser's war effort. With the demand for aero engines minimised after Germany's defeat and the signing of the Treaty of Versailles in 1919, the company was forced to look elsewhere for business, and in 1922 it changed its name to BMW and diversified into building engines for trucks, boats and motor cycles.

From there it was but a small step into building complete motor cycles, and the 1923 shaft-drive R-23 model proved a great success, initiating a tradition of motorcycle excellence from BMW which remains to this day. Looking to expand, BMW then bought the Dixi company of Eisenach, which had just started production of a licence-built Austin Seven under the name of the Dixi 3/15PS. BMW kept this model in production as the BMW 3/15 until 1932, but thereafter the connection with Austin was broken and BMW developed their own cars.

By 1933, the company was sufficiently confident to introduce a completely new six-cylinder model, and the engine of this was gradually developed through the 1930s to power a range of increasingly sophisticated saloons and increasingly potent sports models. Particularly notable were the 326 saloon of 1936, which had an advanced chassis design and a well-proportioned streamlined body, and the 327 coupé and 328 sports roadster, which had tuned versions of the 326's 2-litre engine. All three cars went on to have enormous influence on other European models: in Britain, for example, the 326 influenced the post-War Bristol 401, the 327 influenced the Riley RM, and the 328 influenced the Jaguar XK120.

BMW Before the 2002

By the end of the 1930s, BMW had reached a position from where they could realistically challenge Mercedes in the German car market, although the marque was little known abroad except through its sporting activities. Meanwhile, the company's aero engine division had also been expanding rapidly, favoured by government contracts as Germany rearmed under Hitler's National Socialist regime. When War came in 1939, car production stopped and BMW slipped naturally into the role of full-time aero engine manufacturer. And the engines it produced were among the very best seen on either side during the conflict. Among them were the type 801 radial engine used in early versions of the highly respected Focke-Wolf 190 fighter, and the model 109-003, the first series-production jet engine ever to power an aircraft.

BMW's most remarkable 1930s models were the coupé versions of the **327/28** models.

After the War

BMW found themselves in serious difficulties when the War came to an end in 1945. Their Munich factories had been destroyed by Allied bombing and their Eisenach plant was now in Soviet hands in East Germany. There could be no question of resuming car production immediately, and in fact it would be seven years before another BMW car was made. In Eisenach, meanwhile, pre-War BMW designs were revived as early as 1945 under the EMW (Eisenach Motoren Werke) name, and variants of the 326 and 327 remained in production until 1955.

BMW themselves started production again in a small way, beginning with motorcycles in 1948. Sales boomed, and provided the finance for a new car called the 501, which was introduced in 1952. Yet the 501 was a distinctly odd choice for a car market which was only just recovering from the effects of the 1939-1945 War. It was a large and expensive saloon, which followed on from where the pre-War 326 and 335 had left off, but was competing for a relatively small market with the established Mercedes-Benz 220. When Mercedes announced their

BMW Before the 2002

The **501** saloon was BMW's first post war car. Known as the "Baroque Angel" it was initially powered by a six-cylinder engine and later by a V8. Sales were never strong.

The **503** coupé of the later 1950s used some of the big saloons' running gear, but also failed to re-establish BMW on a sound footing.

The V8 powered **507** sports car was BMW's exotic answer to the Mercedes SL models, but it was decisevely beaten in the market-place by its rivals from Stuttgart.

BMW Before the 2002

new "Ponton" 220 in 1954, the BMW stood no chance at all. New engines - including Europe's first volume-produced V8 - and a variety of improvements over the next 10 years could not establish the big BMW "Baroque Angel" as a major seller. Expensive 503 cabriolet and coupé derivatives and a delicious roadster called the 507 built to rival Mercedes' 300SL also failed to fill the company's coffers, and BMW soon recognised that they would have to look elsewhere for their volume sales.

In choosing a new product area, BMW almost literally went from the sublime to the ridiculous. From 1955, in complete contrast to their luxurious saloons and expensive two-door models, they started making the Isetta bubble-car under licence from Iso in Milan. This little economy car sold nearly 162,000 examples over the next seven years, but it did not compensate for the expensive failures of the 503 and 507 models. Not least of the reasons was that BMW had to pay a handsome royalty to Iso for each one they built.

It was clear that BMW needed to design their own small car, and in 1957 they introduced the BMW 600, a step in the right direction even though it was little more than a stretched and re-engineered Isetta powered by a BMW motorcycle engine. A further step came in 1959 with the 700, again powered by a rear-mounted BMW motorcycle engine but this time looking like a "real" car thanks to handsome three-box styling by the Italian consultant Michelotti. This, too, sold well, but it failed to bring in the big profits which BMW so badly needed.

Not until the 1960s did BMW once again manufacture cars worthy of its own past. Among the first was the technically interesting and neatly styled **700**. This is a 1962 **700LS** model.

The vast gap between the big 501/502 models and the little Isettas, 600s and 700s caused BMW serious image difficulties, and that in turn affected sales. The obvious solution was to introduce a medium-sized car which would both sell in large enough volumes to bring in the money and could serve as a focus for the company's image. Not surprisingly, it needed an exceptional car to fulfil these demands, and BMW spent a lot of time and money in the later 1950s experimenting unsuccessfully with a variety of designs for a medium-sized family saloon.

BMW Before the 2002

The BMW **ISETTA** bubble car was primarily an attempt to generate sales volume in the mid-1950s. It could have hardly been more different from the large luxury saloons and sports cars which were BMW's other products of the time.

This low-angle shot was clearly intended to make the BMW **600** look larger than it really was. Essentially, the car was a redeveloped and enlarged Isetta bubble car.

BMW Before the 2002

Crisis and survival

By 1959, BMW were in a bad way. That year, the company lost 15 million Marks on a turnover of 150 million. A general meeting of shareholders was called during December, and there was strong support for a proposal from the bank which was BMW's chief creditor that the company should sell out. The most likely buyer was BMW's arch-rival, Daimler-Benz.

However, the proposal was blocked by a substantial minority of shareholders, who voted for a counter-proposal to find another source of funding which would enable BMW to remain independent. No doubt a good deal of wheeling and dealing was subsequently done behind the scenes, but over the next few months, two businessmen who already owned a substantial proportion of BMW shares began to increase their holdings. By the autumn of 1960, some two-thirds of BMW shares belonged to the brothers Harald and Herbert Quandt, and BMW once again had funds.

Salvation came in 1961, in the shape of the medium sized saloon which BMW called the **1500**. Within two years, its sucess had turned the company's fortunes around

The Quandt brothers lost no time in getting to work on the company's weak spots. They took on new personnel for some of the key management jobs, and put top priority on the project to develop a family saloon. During 1960, the work already under way was aborted, and new guidelines were set. What was simply known as the "Neue Klasse" (New Class) car would employ no carryover engineering but would be designed from scratch. At its heart would be a brand-new engine of at least 80bhp to be designed by Alex von Falkenhausen. Chassis engineering would fall to Eberhard Wolff and his team; the body and styling would be overseen by Wilhelm Hofmeister; and the project would be co-ordinated by Chief Engineer Fritz Fiedler.

BMW worked quickly. A prototype of the new family saloon was shown at the Frankfurt Motor Show in September 1961, and the car went on sale a year later. Badged simply as a BMW 1500, the new model was immediately acclaimed as an outstanding motor car, and it established at a stroke most of the characteristics still associated with the marque nearly three and a half decades later. Moreover, it sold well, both at home and abroad. By 1963, sales of the 1500 had completely reversed BMW's fortunes, and the company's managers were able to look to the future with confidence.

The 1500 had been enormously expensive to develop, and without funding from the Quandt brothers it could never have been made. It had a completely new unit-construction four-door body shell with smart and attractive styling and adequate room for five passengers within compact overall dimensions. Its 1,499cc four-cylinder overhead-camshaft engine was smooth and powerful, with an almost sporting feel about it which was backed up by a slick four-speed Getrag gearbox. MacPherson strut front suspension and an independent rear suspension with semi-trailing arms gave both good ride comfort and excellent handling, while front disc brakes and a ZF worm-and-roller steering box made their own contributions to a well-balanced package.

Building on success

With the New Class models showing the way ahead and gradually restoring the company to profitability, BMW's senior figures began to plan their next moves. As the New Class had been so costly to put into production, there could be no question just yet of another completely new model. Instead, the company would have to develop further new products from the New Class itself. And over the next few years, that was exactly what BMW did.

The mid-1960s saw BMW building on their new-found success with coupés like this **2000CS**.

Alex von Falkenhausen's four-cylinder engine had plenty of development potential in it, and in 1963 it was given a longer stroke to provide 1,773cc and 90bhp for an 1800 version of the New Class saloon. A year later, the short-stroke crankshaft of the 1500 was mated to the big-bore 1800 block to produce an 83bhp 1,573cc engine, and this went into a 1600 model which replaced the original 1500. Next came a more ambitious project, in which the floorpan and running gear of the New Class saloons were used as the basis of an exclusive coupé called the 2000C, introduced in 1965 with a 1,990cc engine which was the same four-cylinder design with its bore even further enlarged. In standard trim, this had 100bhp, but a twin-carburettor edition with 120bhp was available for the top-

BMW Before the 2002

model 2000CS, and in due course the 2-litre engine would also find its way into the New Class saloons.

These, though, were only the models which made it into production. As early as 1963, there had been discussions about a new and smaller BMW with an engine of around 1.4 litres, and after this had been rejected there were also plans to create a smaller version of the Karmann-bodied coupé. Yet none of these projects seemed to be quite what BMW needed, and so thoughts began to focus instead on a short-wheelbase two-door version of the New Class saloon. This plan had particular merit because the two-door car would help to re-establish the sporting imge which BMW had enjoyed in the 1930s but which it had never recaptured since. The shorter wheelbase would bring handling advantages while the lighter body would improve performance, and of course the two-door configuration would look more sporting than the four-door New Class type.

The two-door model did not take long to design. The New Class wheelbase was shortened from 100.4 inches to 98.4 inches, Wilhelm Hofmeister restyled the passenger cabin to suit, and the front of the car was given a minor facelift. Most of the running gear came directly from the existing four-door saloons, although there was a narrow-track rear axle which made front and rear tracks equal on the two-door model. BMW decided to launch the car with the 1,573cc "1600" engine and, for want of a more inspired name, decided to call it a 1600-2, the additional figure standing for its two doors and distinguishing it from the four-door 1600 saloon.

The 1600-2 was announced in March 1966 and was immediately acclaimed as a winner. The lighter body made the car nearly as fast as the more powerful 1800 saloon, while the excellent handling added a sporting ingredient which was lacking in the larger car. The motoring press were unable to resist comparisons with Alfa Romeo's sporting saloons, and that suited BMW's needs perfectly. At the Frankfurt Motor Show in autumn 1967, they announced an even more sporting version - the 1600ti - with a 105bhp twin-carburettor engine. And by this stage, BMW were already considering the possibility of giving their two-door saloon yet another engine in the shape of a 2-litre relative of the 1600ti's four-cylinder.

BMW Before the 2002

The new two-door models had already been introduced by the time this **2000tii** was built in 1970. Its bodyshell was essentially that of the New Class saloons, but it sported a restyled front end.

BOT 4J

Chapter Two

The early 2002 from '68 to '71

Not long after the 1600-2 was announced, Alex von Falkenhausen had a 2-litre engine dropped into an example of the car for his own use. Completely independently, BMW's Planning Director Helmut Werner Bönsch had exactly the same conversion carried out for his car. Neither man knew of the other's car until one day in mid-1967 when both cars were in the workshops together at BMW. Both were enthusiastic about their 2-litre two-doors, and between them they decided to put a formal proposal to the BMW Board that such a model should be considered for production.

Their cause was greatly helped by developments in the USA. BMW had never meant very much in that market before the mid-1960s, and the marque had been imported only in penny numbers. The company was well aware of the value of sales success in such a large market, but until this point it had not had a model which appealed to American customers. In 1966, all that changed. The new two-door 1600-2 model received rave reviews in the American motoring press, and all of a sudden sales started to gather momentum. Wanting to capitalise on this success, importer Max Hoffmann urged the Bavarians to let him have another model in the same vein, and preferably one with even more performance.

The only model in BMW's range which fitted that description was the 1600ti, introduced a year after the 1600-2 and essentially the same car with a 105bhp twin-carburettor engine in place of the 1600-2's 85bhp single-carburettor type. Unfortunately, the twin-carburettor engine could not be made to meet the new Federal exhaust emissions regulations, and so the 1600ti could not be sold in the USA. However, the 100bhp 2-litre engine in the 2000C coupé *had* been made to meet emissions regulations.

The early 2002 from '68 to '71

The solution was simple. BMW Sales Director Paul Hahnemann was well aware of the US market requirement, and so he supported the proposal for a 2-litre version of the two-door car. Despite opposition from Chief Development Engineer Bernhard Osswald and Production Director Wilhelm Heinrich Gieschen, the sales argument won the day, and the 2002 was born.

There were three distinct "generations" of the BMW 2002 range during its eight and a half year production life. The first generation cars were built between 1968 and 1971. The second generation, or Model 71, cars were built between 1971 and 1973. And the third generation, or Model 73, cars were built between 1973 and the end of production in 1976. Within each of those generations, the 2002 range was further subdivided into a variety of different models. And in the USA, the position was even more complicated: changing legislation there obliged BMW to build Model 74, Model 75 and Model 76 cars, all with certain differences from one another.

The original 2002 (1968-1971)

The BMW 2002 was announced in January 1968, initially for European markets only, but US-market versions started coming off the lines just two months later. In the BMW range, it slotted neatly in between the 1600-2 and the 1600ti, the latter model having a much higher level of standard equipment which justified its higher price.

In order to keep prices down, BMW had been unwilling to make any major changes to distinguish the new 2002 from the 1600-2 which had the same two-door body shell. Badges differed, of course, but the only way of distinguishing a 2002 from a 1600-2 at a glance was from its radiator grille. Where the smaller-engined car had a bright aluminium grille, the 2002's grille had blacked-out slats which gave the car a much more purposeful and sporting appearance.

Inside, the 2002 shared the 1600-2's low levels of standard equipment, and soon came in for criticism as a result. BMW responded swiftly, and from 1st April 1968 - barely three months after the cars had been announced - all 2002s had a sports steering wheel and safety belts as standard equipment. At the same time, a rev counter and reclining front seats could be specified at extra cost to bring the interior up to 1600ti levels. To complete this first package of revisions, the 2002 was fitted as standard with anti-roll bars front and rear. Not surprisingly, its basic showroom price increased to cover this extra equipment, but the differential between 2002 and 1600ti was maintained because the price of the latter model was increased at the same time.

The early 2002 from ’68 to ’71

The **2002** was announced in 1968. This is an original press photograph of the time.

The early 2002 from '68 to '71

The early 2002 from '68 to '71

In Britain the **2002** was not a strong seller before 1971. The whitewall tyres would probably not have been supplied when the car was new, and the door mirror and wiper blade spoiler aren't original items, but in all other aspects this well-preserved example clearly illustrates the essential features of the early **2002**.

The early 2002 from '68 to '71

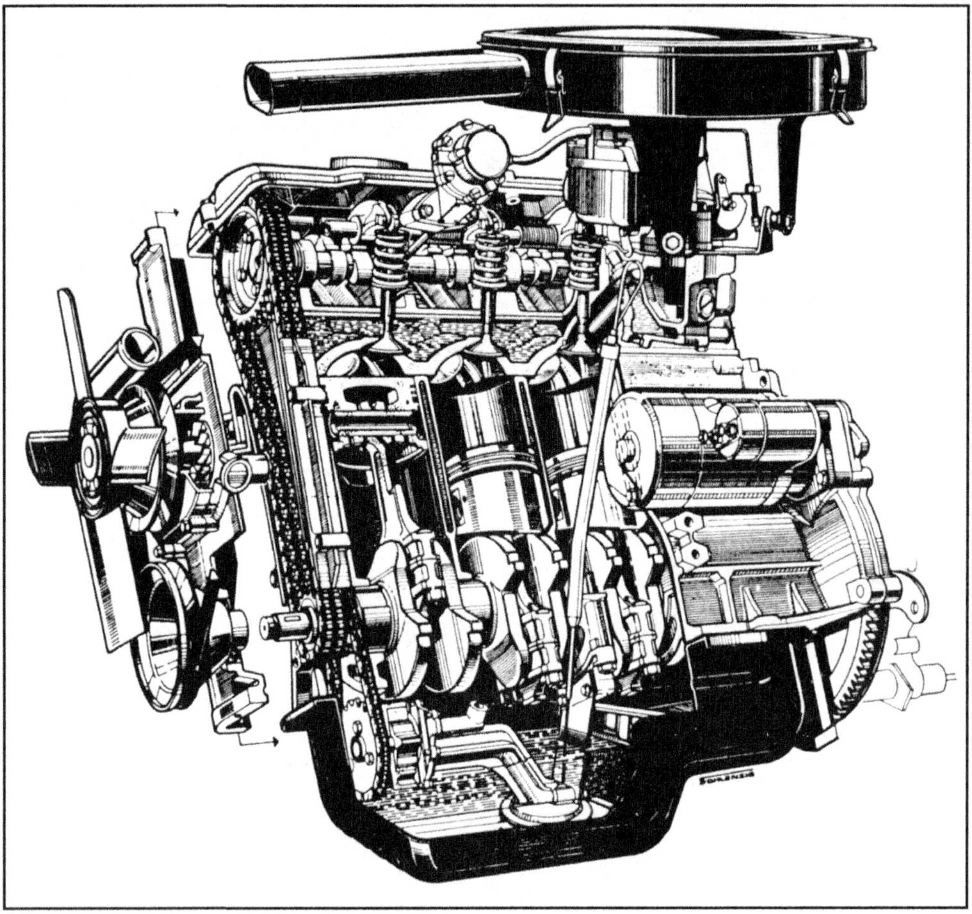

The 2-litre engine in the **2002** was a real gem. This cutaway illustration shows the original version of the engine.

The early 2002 from '68 to '71

The key ingredient in the 2002 was its power train, of course. The 2-litre type 121 engine was a bored and stroked derivative of the 1.6-litre used in the 1600-2 and 1600ti and, like its smaller-capacity cousins, it was installed in the two-door shell with a slant of 30 degrees to the right in order to fit under the low bonnet. Also like the 1.6-litre "four", it had a cast-iron block with an aluminium alloy cylinder head. Combustion chambers were described by BMW as "twin hemispherical swirl action" type, designed to give the incoming fuel-and-air mixture a turbulence which improved combustion efficiency. The overhead camshaft was driven by a double roller chain and operated both inlet and exhaust valves by means of rockers, and the crankshaft ran in five main bearings.

With 100bhp at 5,500rpm in European trim, the type 121 engine gave the 2002 a maximum speed of around 107mph. This in itself was exciting enough for a compact 2-litre saloon in the late 1960s, but what really distinguished the BMW from its rivals - and from the more expensive 1600ti - was the high torque output of that engine. It allowed tall final drive gearing of 3.64:1 to be fitted without detriment to the in-gear acceleration, and that high gearing made for relaxed high-speed cruising and quite excellent touring fuel economy. The careful matching of overall gearing to the engine's torque characteristics made sure that the car could be driven easily in relaxed mode, or that very rapid acceleration was available on demand to the more sporting driver who was prepared to use the gearbox to get the best out of the car.

The standard gearbox was the four-speed already familiar from other BMWs, and until July 1968 it was the only one available. In that month, a five-speed close-ratio box was announced as an option for both 2002 and 1600ti, but it was too expensive to attract many customers. However, it did add a certain zest to 2002 acceleration and was very much appreciated by more sporting drivers. Later modifications to the car were few, although a dual-circuit braking system was added in autumn 1968.

The 2002ti (1968-1971)

The sales success of the 2002 was every bit as good as BMW could have wished for, and they were no doubt delighted to discover its sales were incremental: in other words, the 2002 was opening up a new market for BMW without eating into sales of their existing models. The next stage, inevitably, was to exploit the concept further.

BMW had of course started to look beyond the initial 100bhp 2002 model even before that car hit the market. If the single-

The early 2002 from '68 to '71

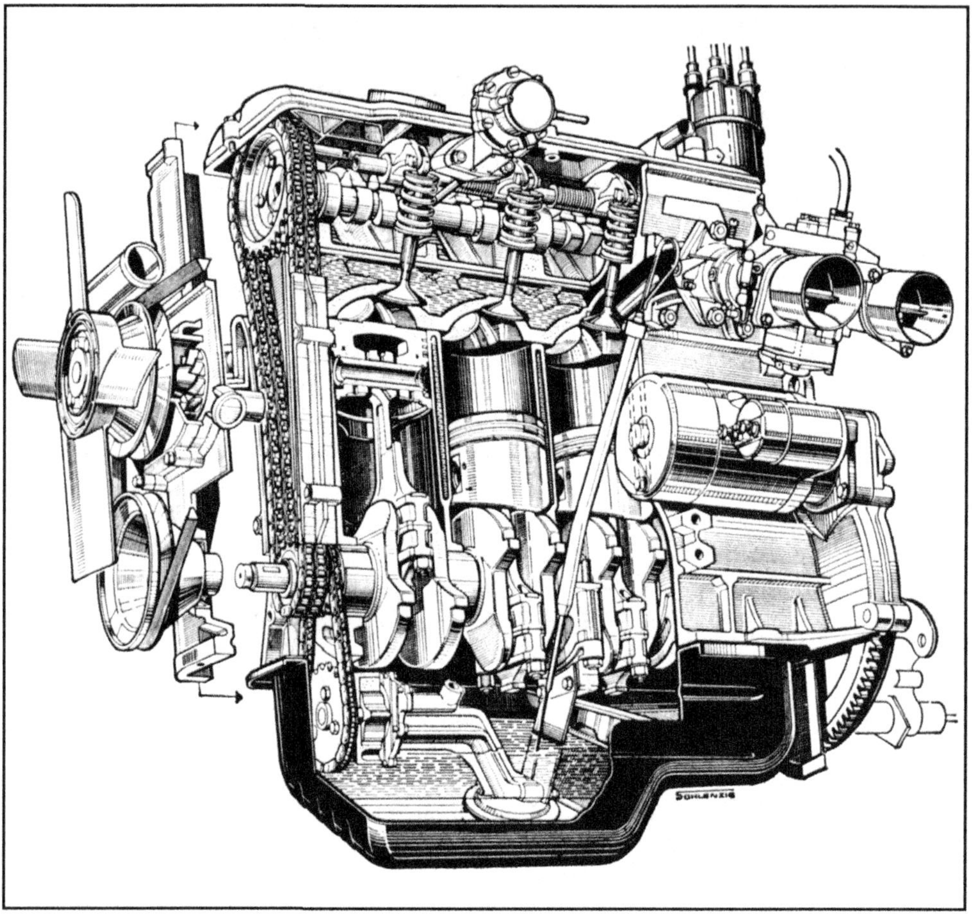

The **2002ti** twin-carburated engine.

carburettor 2-litre engine could be made to fit the two-door shell, there was no reason why the 120bhp twin-carburettor version of the engine should not be made to fit as well. This engine was already available in the 2000ti lux four-door models and in the 2000CS coupés, and so development work would be minimal.

By March 1968, just two months after the launch of the original 2002, BMW were confident that a twin-carburettor version was a viable proposition both technically and in terms of market demand. That month, they announced that there would in due course be a twin-carburettor car called the 2002ti. The actual launch nevertheless did not take place until October, mainly because the production lines were already at full stretch to cope with demand for the standard 2002 and no spare production capacity was available. It was primarily for that same reason that no right-hand drive versions of the 2002ti were ever made.

The **2002ti** was distinguished from the outside only by its badging; in all other respects it looked like an ordinary **2002**. This example has wider-than-standard wheels and tyres.

The production 2002ti was rather more than a twin-carburettor edition of the original car, however. To cope with the additional power and torque, it had box-section rear suspension swinging arms. Wider wheels with 5-inch rims instead of the 4.5-inch rims on the standard 2002 gave a wider track which improved the handling, while there was also better stopping power from larger brakes. The front discs were both larger in diameter and fatter than those on the standard 2002, to give an increase in swept area of no less than 37%.

These changes had certainly not been made simply to enhance the paper specification of the 2002ti. The 120bhp engine made the car much quicker than the single-carburettor model, putting the top speed up from 107mph to 118mph, which was quite astonishing for a 2-litre saloon in the late 1960s, especially one as unstreamlined as the 2002. Without the improved handling and uprated brakes, the 2002ti would have been good for straight-line acceleration but probably less than fully reassuring when it came to cornering and stopping. As it was, the model retained the balance of dynamic qualities which had made the original 2002 such an excellent piece of machinery, while adding in an exhilarating extra dose of performance.

The early 2002 from '68 to '71

As the 2002ti did not become available until the autumn of 1968, all production examples had the new dual-circuit braking system, with four-pot front calipers. The ti also benefited from a hazard warning light system and an uprated alternator which were not available on the ordinary 2002. A larger-capacity radiator took care of the greater cooling demands of the 120bhp engine, and there were new flexible joints on the propshaft to suppress any driveline vibration. Only sharp eyes would have spotted the fatter tyres and different rear badging which distinguished 2002ti from 2002, but there was also one readily-visible give-away in the shape of a "2002ti" script badge on the grille.

The 2002 Automatic (1969-1972)

Just as the 2002 had been created by mixing and matching existing components - the 1600-2 bodyshell and the 2-litre engine - so the 2002ti had been created by adding the engine and suspension components of the 2000ti saloon to the 2002. Once again borrowing existing components from other BMW models, the company created a third 2002 variant in the shape of the 2002 Automatic. It went on sale in January 1969.

This time, the componentry came from the 2000 Automatic saloon, and took the form of the complete powertrain with its 100bhp 2-litre engine and ZF three-speed automatic transmission. Linked to the same 3.64:1 final drive as the single-carburettor 2002, this created a variant which was rather slower than the manual-transmission cars, although it was still fast enough to qualify as a quick saloon when new. Zero to 60mph took around 13 seconds, which was some 3.5 seconds more than a manual 2002; and of course top speed was also down, at around 102mph. Yet these were figures not to be sniffed at, and the automatic 2002 still drove like a responsive sports saloon.

There has been some suggestion that the automatic transmission was introduced primarily to appeal to customers in the USA. It certainly did find its own market, both in the USA and elsewhere, but as the production figures in Appendix C show, it was never a very strong seller.

The 2002 Cabriolet (1971)

BMW Sales Director Paul Hahnemann had been in favour of developing the basic four-door New Class models by means of body and engine variants since the middle of the 1960s, and without his support there might never have been a two-door body shell. Not surprisingly, he also lent his support to a proposal for a convertible derivative of the New Class.

The early 2002 from '68 to '71

The first design for such a model appears to have been drawn up in-house by BMW's styling chief Wilhelm Hofmeister. However, Hofmeister wanted to build a completely new bodyshell on the floorpan and running gear of the 1600 saloon, and his proposal was ruled out as being too costly. So BMW showed interest when Karosserie Baur of Osnabrück, who were already building the bodies for the big 2000 coupés, offered to draw up a convertible version of the two-door "02" body.

The cabriolet derivative by Baur was available as either a **1602** or a **2002**, badging being the only external difference. Visible at the bottom edge of the front wing is the coachbuilder's own badge.

Seizing their opportunity, Baur also drew up plans for a fixed-head coupé version of the two-door shell, but BMW decided not to proceed with this after examining the prototype. The full convertible, or Cabriolet, nevertheless scored a hit with BMW management. Baur's prototype was built from a 1600-2, and had a shortened rear end which was rejected, most probably on the grounds of cost. A modified version, using standard rear end panels, was displayed at the Frankfurt Motor Show in autumn 1967, and the Cabriolet went into production in 1600 form in January 1968. Karmann's two-seater 1600 cabriolet, shown on the coachbuilder's stand at that same Frankfurt Show, never stood a real chance against the Baur cabriolet and remained a one-off.

With the 2002 well established as a range, and with range expansion under way, BMW decided to use the cabriolet bodyshell as the basis of yet another 2002 variant. So it was that production of the 2002 Cabriolet, with the 100bhp single-carburettor engine and manual transmission, began in January 1971. By this time, however, Baur must already have been working on the new Targa bodyshell for the 02 models, and it is therefore quite surprising that BMW did not wait until the new shell came on-stream that summer before introducing the open 2002. Perhaps the real reason was that there was sufficient demand for an open 2002 to persuade the company to get one into production as soon as they could.

The new Targa bodyshell entered production in June 1971, and the 2002 Cabriolet was therefore only available for a period of around six months. In that time, just 200 examples were built, with the result that the Cabriolet remains the rarest of the 2002 range. All of them had left-hand drive, and none were produced to US specification. All had the cosmetic details of cars built before the April 1971 facelift.

The early 2002 from '68 to '71

The 1968-1971 US models

During 1966, the US Government announced that it would be introducing new Federal Safety Standards with which all cars sold in the USA after summer 1968 would have to comply. These regulations were primarily designed to ensure that a car offered its occupants a good degree of protection in a crash, and covered a whole range of matters from the strength of the structure to the compulsory fitting of safety belts. BMW had no difficulty in adapting the 2002 to meet these regulations because the car's monocoque structure was already inherently strong; but the need to make its engines conform to the exhaust emissions regulations which were imposed on new cars after the summer of 1969 did cause some complications.

The aim of the exhaust emissions regulations was to minimise the amount of noxious gases pumped into the air from vehicle exhausts. To meet the regulations, motor manufacturers devised a variety of different methods, including the one which BMW adopted. This was known as Exhaust Gas Recirculation, or EGR for short. The type 121 2-litre engines equipped with EGR systems had a belt-driven air pump which forced air into the exhaust manifold. This air mixed with the exhaust gases to give what appeared to be cleaner air at the tailpipe - although in fact exactly the same quantities of pollutants were being pumped out! While this met the US Federal regulations, it did nothing for the output of the engine. The air pump itself absorbed some power, while the detuning necessary to allow the engine to run on 91-octane low-lead fuel took much of the edge off the US-model 2002's performance.

The US-model 2002 also had some other differences from the European models. To meet Federal Safety Standards legislation, it was equipped with a dashboard warning light which was activated if the fluid level dropped in one of the hydraulic brake circuits; and it had a warning buzzer and dashboard-mounted lamp to indicate if a front seat had been occupied but its safety belt had not been fastened. Finally, US road testers and customers were initially disappointed not to be able to buy a 2002 with factory-installed air conditioning. Such a system did become available later, however, and one manufactured by Behr in Germany became optional equipment. Several 2002 models sold in the USA were also fitted by the importers with a Frigiking system.

The US market was offered both the manual transmission 2002 (from March 1968) and the automatic model (from May 1969). Unfortunately for North American BMW enthusiasts, however, it proved impossible to make the twin-carburettor engine meet Federal exhaust emissions regulations. For that reason, the 2002ti was never available on the US market.

The early 2002 from '68 to '71

Bertone's **Garmisch** prototype was based on a **2002ti**, and undoubtedly influenced the styling of the first **5-series** saloons in the early 1970s.

The Garmisch Prototype

The Italian styling house of Bertone had designed the bodywork of BMW's big coupés in the early 1960s, and Nuccio Bertone had also acted as a consultant for the styling of the 1500 New Class saloons on which the 02 models were based. In 1969, Bertone styled a new saloon, this time on the basis of a 2002ti.

A single prototype was built and was displayed at the Geneva Show in March 1970, wearing 2002ti badges and bearing the name of "Garmisch". Although the wheelbase was identical to that of a 2002, the car was actually 4cm (1.57 inches) wider and 12cm (4.7 inches) longer, the extra length going into a larger boot. It was also 9cm (3.5 inches) lower overall, which reduced headroom considerably. Like the production cars, the Garmisch prototype was a two-door four-seater.

The car's glass area was much larger than that of production 2002s, with steeply raked front and rear screens. To protect the rear passengers from the sun, the rear window was covered by a curious honeycomb panel, which nevertheless allowed the driver to see the road behind. The interior included a futuristically styled dashboard and steering wheel, and a combined writing and make-up table pulled out of the passenger side of the facia.

Although no production 2002 was influenced by the Garmisch prototype, the exterior styling certainly did have an impact on the four-door 5-series saloons introduced in 1972.

The early 2002 from '68 to '71

The BMW Diana

After his career with the BMW works team was over, Hubert Hahne set up as a BMW dealer in Düsseldorf. He had bought his wife, the actress Diana Körner, a specially-equipped 2002 as a wedding present, and this car attracted considerable attention. This persuaded Hahne to create a luxury edition of the 2002ti, with BMW's permission, which he named the BMW Diana.

There were just 12 of these cars, numbered 1 690 201 to 1 690 212. To enhance their appeal, Hahne had each one finished a different colour, and none of these colours figured on the BMW standard options list. In addition, the interiors were all different. The conversion work was carried out by Baur in Stuttgart. BMW Diana specifications included:

Hubert Hahne with his wife, the actress Diana Körner.

Opposite page: the car she inspired the BMW **DIANA**

electric sunroof
leather upholstery with Recaro front seats
black vinyl headlining
velour carpet and boot lining
leather binding for the carpets
black vinyl lining for the boot lid
leather gaiters for gear lever and handbrake
chromed handbrake lever
aluminium tread-plates on the sills
front and rear overriders
Becker stereo radio
laminated windscreen
heated rear window
ammeter and clock
special silver grille
twin headlamps (from the BMW 2800)
chromed covers for the bumper bolts
special exhaust
Sebring-type driver's door mirror.

Hahne introduced the BMW Diana in spring 1971, at a price of 22,500 DM; a standard 2002ti then cost 12,321DM. The first five examples sold quickly, but the other seven went more slowly, the last example being sold at the end of 1971. To encourage sales, Hahne offered to equip them with injected engines or turbochargers.

Hahne retained one car himself - finished in Coffee Brown. Into this he had the tuning specialist Jürgen Grähser transplant the 150bhp 2.5-litre six-cylinder engine from a BMW 2500 saloon. The conversion also included a five-speed gearbox, limited-slip differential and electric windows.

The early 2002 from '68 to '71

D TU560

Chapter Three

The 2002 from '71 to '76

The 2002 and 2002 Automatic (1971-1973)

The 2002's sales success could not go unchallenged for long. Just eighteen months after the original 2002 went on sale, buyers in the market for a sporting four-seater car were offered a cheaper alternative in the shape of Ford's new Capri; and a year after that, Opel responded to the challenge with the Manta. By the end of 1970, the 2002 was therefore being threatened by these two newcomers, and BMW decided it was necessary to revitalise market interest in their two-door sports saloons. To do so, they came up with a series of revisions, a facelift which turned the cars into what are now known as the "Model 71" types. These facelifted models went into production during April 1971 and were available in the showrooms during May.

Externally, the Model 71 cars were easy enough to distinguish from their forebears. The rear bumper now wrapped further around the sides of the car, reaching almost as far forward as the wheelarch, and the sides of the body wore bright bump-strip mouldings with black rubber inserts. These were matched by rubber facings on the bumpers and overriders.

The passenger compartment had also been thoroughly overhauled. Reclining seats were now standard on all models, and improved upholstery gave better occupant location. The dashboard itself had lost the brightwork which had dated it so obviously to the 1960s, and the instruments were now easier to read than before. There were new tell-tale lights for the choke and handbrake, and a low-fuel warning light had been added. The screenwash now incorporated a wash/wipe cycle, and its control switch had been removed from the dashboard and repositioned more accessibly on the steering column. Under the instruments, the foam knee-pad first introduced on US-market models was now standard, while a three-spoke steering wheel replaced the

The early 2002 from '71 to '76

earlier two-spoke type. The horn push was now in the centre of the wheel, and operated two-tone horns instead of the single-tone type of earlier 2002s. There was new ribbing on the passenger side of the dashboard, to prevent oddments from sliding about; and the heater output had been increased by 20%.

Lastly, there were some less visible changes. All models now had flexible propshaft joints, a hydraulically operated clutch, a more powerful 630 Watt alternator, and a five-blade engine cooling fan. In addition, Borg Warner synchromesh replaced the Porsche type of earlier gearboxes, giving a slightly different feel to the gearchange.

The Model 71 2002 and 2002 Automatic continued in production essentially unchanged until they were replaced by the Model 73 types in mid-1973. However, cars built after 1st September 1972 had a slightly different engine known as the E12 type. Identical in all its essentials to the older 121 engine, it had what BMW called 'triple hemispherical combustion chambers', which had first been seen on the six-cylinder engines and improved combustion so that the engines would meet US emissions regulations. To streamline production, the new combustion chambers were introduced for all models and all markets at the same time.

The frontal aspect of the Model 71 cars is typified by this early right-hand drive **2002tii**. Note in particular the rubber bumper facings, the black grille bars and the rubbing-strips along the car's flanks.

The 2002 tii (1971-1973)

When the facelifted Model 71 cars were introduced in the spring of 1971, the twin-carburettor 2002ti went out of production and was replaced in the 2002 range by a new model known as the 2002tii. The second "i" in its designation stood for "injection", for this new model had the fuel-injected engine from the four-door 2000tii introduced in 1969. In 1971, fuel injected engines were still quite rare, and they brought with them an aura of high technology, power and performance.

The early 2002 from '71 to '76

That was precisely what BMW wanted in order to keep their 2002 models ahead of the pack of Ford Capris and Opel Mantas which were now snapping at the 02 saloons' heels.

Back in 1969, when the 2000tii had been introduced at the Frankfurt Motor Show, BMW had also shown a prototype 2002tii. However, the car would not enter production for a further two years, either because further development work proved necessary or because BMW simply could not afford to introduce yet another new model at that stage. Then in April 1971, a quirk of marketing policy caused the Touring version of the injected 2-litre 02 to be announced a few days before the saloon version. Finally, another quirk of policy at BMW caused the Touring model to be badged initially as a Touring 2000tii. Not until 20th April 1971 did the 2002tii saloon proper see the light of day.

The greatest advantage of the 2002tii as far as BMW were concerned was that it could be tuned to meet US exhaust emissions regulations, and therefore gave the company another model to sell in that booming market. For owners, however, the advantages were quite different. Foremost among them was undoubtedly the extra performance available from the injected engine, which had 130bhp and promised 0-60mph acceleration times of under 9 seconds with a top speed of 118mph. But by no means negligible were two other advantages: first, the injected engine actually gave better fuel economy than the lower-powered single-carburettor engine in the standard 2002; and second, it eliminated the awkward tuning and balancing of the twin-carburettor installation on the superseded 2002ti.

This rather later car, again a **2002tii** but this time wearing optional Mahle alloy wheels, shows the wraparound rear bumper which was introduced with the Model 71 cars.

However, the 2002tii was not quite all it appeared on paper. At everyday town speeds, it was barely quicker than a single-carburettor 2002, and most of its extra performance was available only if the engine was revved quite hard (maximum torque was at 4,500rpm as compared to 3,000rpm in the single-carb engine). For all practical purposes, then, the 2002tii was unable to demonstrate the merits of its injected engine except in long-distance, high-speed motoring.

Naturally, the 2002tii was distinguished from other 02s by small details. Like the 2002ti, it had a rev counter as standard, but it also had

The early 2002 from ’71 to ’76

HR-rated tyres and a leather-rimmed sports steering wheel - although UK-market cars seem not to have had the special steering wheel. Unique to the tii was a 230kph (or 140mph) speedometer, which replaced the standard 190kph (or 120mph) type. Unlike the superseded ti, however, the injected 2002 had no badge on its radiator grille, and was therefore harder to distinguish from the standard model at a distance.

It is interesting to speculate whether BMW would have introduced the 2002tii with Bosch electronic fuel injection if the funds had been available. The six-cylinder engines introduced in other models in spring 1971 had this system, which was very much more advanced than the Kugelfischer mechanical system which had always been used on the 2-litre engine. However, at the turn of the decade, BMW was investing so heavily in new models that to have developed a four-cylinder engine with the Bosch injection system would have been little more than an additional headache. The Kugelfischer system, outmoded though it was, therefore remained in the specification; and it would remain so until the last 2002tii came off the production lines in October 1975.

Opposite page: The **2002tii** had a fuel-injected version of the 2-litre engine.

This page: The **2002tii** engine installed. Note the cylindrical air filter on the side nearer the camera.

Nevertheless, BMW did make some changes to the engine during the 2002tii's production run, and those after the car had been on the market for no more than 12 months. Starting difficulties with the original type 121 engine were largely cured by a new cylinder head and pistons, and by a modification to the injector pump; the revised engine was known as the E12 type. Power outputs remained unchanged, although the later engines are often wrongly accused of having rather less bite than the originals.

The 2002 Targa (1971-1973)

BMW showed their new Targa-top 2002 at the Amsterdam Motor Show in the early spring of 1971, and the model went into production that June. It replaced the full convertible or Cabriolet version of the 2002 which had been introduced just six months earlier, in January 1971.

There were two main reasons why the full convertible passed into history. The first of these was that its bodyshell was not as rigid as that of the saloons, and BMW did have some worries about additional vibration induced by the more powerful 2-litre engines. The second reason was that US legislation was widely expected to ban open cars altogether on safety grounds in the next few years. Many motor

The early 2002 from '71 to '76

The early 2002 from '71 to '76

Left:
The very earliest targa top models were converted from pre-Model 71 cars as this example shows. However, there were probably very few cars built to this spec; the production cars were Model 71 and later types.

Above:
2002 TARGA models built in 1971 and 1972 had their rollover bars painted in the body colour. It looked good when the soft top was down, but rather odd when it was raised.

manufacturers who wanted to sell cars in the USA deliberately avoided building new convertibles for exactly that reason in the mid-1970s, or built open cars with rollover protection. Typical of the times was the British Triumph Stag of 1970, introduced with a massive T-bar to protect its occupants in a rollover accident.

Rather than abandon the open 2002 altogether, BMW decided to give it appropriate rollover protection. In conjunction with the Stuttgart coachbuilder Baur, who was already building the 2002 Cabriolet bodies, they came up with a stylish design in which the rollover bar was blended into the body sides just behind the rear doors. To help improve its appearance, the rollover bar contained opening quarter-light windows on each side. Behind it, a half-hood incorporated the rear window and folded down into a recess behind the seats. The main roof panel over the front seats, meanwhile, could be lifted out for open-air motoring and stowed in the boot on special brackets, where there was still room for luggage below. Following the example of Porsche, who had introduced a lift-off roof panel with a fixed rear window as an option for their 911 in 1965, BMW decided to call this new model a Targa.

There was no doubting that this arrangement made for a very distinctive car, and one which was torsionally very much stiffer than the earlier full Cabriolet. Sales were never enormous, partly because the car was very expensive, although their average of 650 a year compared well with the 200 full Cabriolets built in just six months of production (a theoretical 400 a year). According to factory records, all the Targa-top cars were built with the 100bhp 2-litre engine with manual transmission, and there were no automatics or tii versions. Very early left-hand drive cars had the rollover hoop painted in the body colour, but black vinyl coverings were added early in production to improve the appearance with the hood up.

The Touring models (1971-1973)

A third body style for the BMW 2002 range was announced in April 1971, to join the existing saloon and Cabriolet types. This was the Touring, which was made available with both the standard 100bhp single-carburettor engine and the 130bhp fuel-injected tii engine. Confusingly, BMW decided not to call these cars by the 2002 designation but instead to badge them as the Touring 2000 and Touring 2000tii. Not until December 1972 were the model-designations changed to reflect the cars' obvious place within the 02 family, when they became the Touring 2002 and Touring 2002tii respectively. BMW have never explained why the 2000 designation was used in the beginning, although it seems possible that they originally considered that the car could not be called a 2002 because the second "2" of that number had originally stood for "two doors" and the Touring had three! Then, of course, the designation "2003" might have been logical ....

The late 1960s and early 1970s saw a number of attempts to build cars which would combine the practicality and versatility of an estate car with the driving appeal of a sports car. The trend had actually begun with Renault's ground-breaking 16 of 1965, which offered little in the way of sporting qualities but did point the way forward. The concept of the hatchback saloon prospered, but few of the attempts to make sporting estate cars sold in large volumes. Reliant's 1968 Scimitar GTE set the trend, which was quickly copied by Volvo with the 1800ES of 1971, but the market sector collapsed when the first hot hatchbacks appeared in the mid-1970s. BMW's Touring, selling in the same market sector, did not sell in large enough quantities to persuade its makers to build a successor. The E30 3-series Touring which appeared in the late 1980s was of course a more conventional estate car, despite the similarity of name.

Early examples wore the **2000 TOURING** badges. The change to **2002 TOURING** was made in December '72.

The concept of the BMW Touring had originated some time around 1967, after BMW had bought out the Glas company. Among the Glas models which were not rebadged as BMWs was a rather ungainly-looking 1,289cc hatchback called the 1304GL. Never a strong seller, the car nonetheless seemed to BMW to incoporate a number of good ideas. Encouraged by Sales Director Paul Hahnemann, the company therefore began to look at the possibility of building a similar style of vehicle on the basis of the existing 02 models.

As BMW's own stylists were heavily committed on other projects, Chief Development Engineer Bernhard Osswald commissioned proposals from two Italian styling consultants. One was Bertone, who had earlier shown some interest in styling bodies for BMW; the other was Michelotti, who had styled the small BMW 700-series cars at the end of the 1950s. Their brief was to design what BMW described as a "Combi-Coupé", or sporting estate.

The early 2002 from '71 to '76

The **TOURING** was essentially a three-door hatchback model. Despite its immense practicality, it was never really a great success.

The early 2002 from '71 to '76

Michelotti's proposal was sufficiently interesting for BMW to commission a prototype, which was running by the middle of 1969 when it was spotted and photographed by the German magazine Auto, Motor und Sport. It was then badged as a Glas, and was wearing the 1304 model badges of the car which had been the original inspiration behind the Combi-Coupé project, although at BMW the car was known as the City. By this stage, however, the project seems to have been on hold because of the runaway success of the 2002 saloons: BMW simply could not commit themselves to building another new model until they had met demand for their existing products.

Nevertheless, BMW's dealers pressed them to put the car into production. Envious of the huge sales of the Ford Capri and Opel Manta, they argued that BMW could and should attempt to compete in this new market for a practical car with sporting overtones. So the Michelotti-styled proposal entered production in 1971 as the BMW Touring, with a choice of 1600, 1800, 2000 and 2000tii powertrains - though only the 2000 was available with right-hand drive.

The Touring used recognisable 02 styling elements, and in fact incorporated the complete 02 front end panelling. However, its windscreen was more steeply raked to give a more sporting appearance and the rear of the body sloped downwards in "fastback" style to a truncated rear end: overall, the Touring was eight inches shorter than an 02 saloon. The rear panel was hinged at the roof to give access to the luggage area in the rear, and the back seats could be folded down individually, estate-car style, to provide a larger load area. Interior appointments were otherwise the same as those of the equivalent saloons.

One of BMW's major concerns when the Touring was introduced was that the car should not be seen simply as a practical holdall which might detract from the marque's sporting image. For this reason, it was introduced in 2000tii version first (even before the 2002tii saloons were announced), and was made available with optional alloy wheels and sports seats which were not available on the 02 saloons. Lesser models followed very soon after the 2000tii, but BMW's marketing point had been made.

The Touring 2000 and Touring 2000tii entered production together in April 1971, although production of right-hand drive cars did not begin until September. An automatic version was introduced in May 1971 but sold poorly and was never made available with right-hand drive. It was withdrawn after only 17 months in production.

The early 2002 from '71 to '76

US models (1971-1973)

Between 1971 and 1973, US customers were offered a restricted range of 2002 models. No Cabriolets, Targas or Tourings were ever sold in the USA (although some may have entered the country as personal imports), and the only models available were the standard 2002 saloon, the 2002 Automatic, and the 2002tii. Although production of the US-variant Model 71 2002 with manual transmission started in April 1971 and that of the 2002tii in August that year, the automatics did not enter production until January 1972. They remained by far the least popular of the models sold in the USA.

To meet emissions control regulations, all 2002 and 2002tii engines built for the USA in this period had both EGR systems and lower compression ratios which allowed them to run on low-lead fuel. As a result, they were less powerful and less accelerative than their standard equivalents. Nevertheless, as all cars sold in the USA were suffering from the effects of the same legislation, the BMWs continued to maintain a performance edge over their rivals, while their excellent handling was of course unharmed by the emissions control changes.

European 2002s, (1973-1975)

At the Frankfurt Motor Show in autumn 1973, BMW introduced a second major facelift for the 2002 range, and at the same time announced one additional new model in the shape of the 2002 turbo. Older models which remained in production for the 1974 season (the Model 73 cars) were the 2002, 2002 Automatic, 2002 Targa, 2002tii, and the Touring 2002 and 2002tii.

The Model 73 cars were distinguished by a new radiator grille, this time a three-piece black plastic moulding with a bright surround to the BMW kidney-grille section in the centre. At the rear, the distinctive round rear lights had been replaced by rectangular light units, and the BMW roundel had moved from the rear panel to the centre of the bootlid edge. However, the Touring models retained their original round rear lights. All cars had new eight-spoke steel wheels, and all models had matt black wiper arms and blade holders in place of the bright metal type fitted to earlier 2002s.

There were interior changes, too. All models had new and more comfortable front seats, borrowed from the new 520 model introduced a year earlier, there were one-piece moulded carpets, and the indicator stalk had moved from the right to the left of the steering column to meet ISO (International Standards Organisation) standards. At extra cost, buyers could order the Lux interior pack for any model of 2002

The early 2002 from ’71 to ’76

Top:
From the front, the black plastic grille was the most obvious feature of the Model 73 cars.

Bottom:
The engine installation was always neat on the **2002**. This is the single-carburettor engine in the 2002 pictured above.

The early 2002 from '71 to '76

From the rear the Model-73 cars were distinguised by their new rectangular rear lights - which were nevertheless not fitted to the Touring models. Note the 'L' which underlines the model name. This indicates a Lux model.

except the Touring and the new Turbo. This brought different upholstery (with cord centre panels and matching velour outer panels on the seats), a fold-down rear centre arm rest, pockets on the door trims, wooden door cappings and a wooden panel behind the ashtray to match the instrument surround on the main binnacle. Also part of the pack were front seat head restraints and a leather-rimmed steering wheel, a dipping rear mirror, inertia-reel front safety belts, an intermittent wash-wipe and a heated rear window. Cars with the Lux pack had special badging at the rear, where the 2002 script was fitted onto a black panel with the letter "L" in bright script at its left-hand edge, its elongated base underlining the rest of the badge.

Lastly, the Model 73 cars also incorporated some more functional improvements. All engines were fitted with a rev limiter on the rotor arm which prevented crankshaft speeds greater than 6,400rpm; the fuel tank had a capacity of 51 litres instead of the earlier 45; and the headlamps were now H4 halogen types.

Continued poor sales of the Touring models persuaded BMW to take them out of production in April 1974. Right-hand drive Targa models had been withdrawn even earlier than that, the last one being built in January 1974. By now the whole 02 range was coming to the end of its production life; work was well advanced on the 3-series range which would replace it, and the first examples were spotted out on test in the spring of 1974. However, there would be one further change to the 2002 models before they were taken out of production: cars built after February 1975 had new front seats, which proved in due course to be those designed for the 3-series.

The 3-series cars were announced in June 1975, and production of the 2002 range was brought to an end in stages after that. Left-hand drive 2002s, 2002 Automatics and 2002tii models actually stopped production that month; next to go were their right-hand drive counterparts, in October. The last left-hand drive 2002 Targas were built in December - the delay being accounted for by stocks of bodyshells awaiting conversion at the Baur factory in Osnabrück - and only the US-specification cars remained in production until they, too, disappeared during 1976.

The early 2002 from '71 to '76

Top left and left: Lux trim became available with the model -73 cars. Trim included wood behind the ashtray in the centre console and on the door panels.

Bottom left: Optional BMW wheels produced by Mahle for the 2002.

The early 2002 from '71 to '76

The **2002 TURBO** engine was a derivative of the standard 2-litre four cylinder type in other **2002s**.

The 2002 Turbo (1973-1974)

Seldom has a car created as much controversy at its launch as the BMW 2002 turbo, and seldom has bad luck so seriously affected a new model's chances. It was extremely unfortunate that the fastest 2002 derivative of all should be introduced to the public only a matter of weeks before the oil crisis struck and turned powerful cars almost overnight into anti-social nuisances. And it was also extremely unfortunate that the 2002 turbo's aggressive appearance should have incurred the public wrath of an influential safety and speed lobby in the car's native Germany.

By the end of the 1960s, BMW already knew that the 2-litre engine had reached the limit of its development potential. Fuel injection was already in the offing for the 1971 2002tii, but it was not possible to boost the engine's power significantly and reliably beyond the 130bhp which the injected engine developed. To extract more power from it would mean overboring the block to gain extra capacity, and this would have resulted in the siamesed cylinder bores to which chief engine designer Alex von Falkenhausen was implacably opposed. As the latest BMW six-cylinder engine was too big to go into the 2002's engine bay, there seemed no way forward.

This was why BMW turned to the unconventional solution of turbocharging for the works 2002 racers which ran in Group Five events during 1969. Turbochargers were then new for petrol engines, although they had been common on large diesels for some time, and the technology was largely untried. However, the 270bhp works racers were sufficiently successful to inspire von Falkenhausen to begin work on a roadgoing turbocharged engine for the 2002 - and to inspire the Swiss tuner Michael May to build his own 170bhp turbocharged 2002 during 1970.

However, turning race-bred performance technology into something suitable for everyday use did not prove easy, and the roadgoing BMW turbo was not ready for production until the middle of 1973. By using a KKK turbocharger coupled to the 2002tii's Kugelfischer mechanical fuel injection, von Falkenhausen had managed to create a 170bhp engine which gave a 2002 saloon the performance of a Porsche 911: 0-60mph in seven seconds and a top speed of 131mph. For its time, the car was a revelation.

Unfortunately, the turbocharged engine could hardly be described as docile. At low revs, it behaved very much like a detuned version of the injected 2-litre engine - which indeed it was, having a very much lower compression ratio than the tii engine and a version of the same Kugelfischer fuel injection. Once enough exhaust gases were being produced to get the turbocharger spinning, however, the additional boost came in with a bang and the car would rocket forward. These were not characteristics which made it easily driveable either in stop-start traffic or on wet and slippery roads.

To cope with the extra stresses imposed by the additional performance, BMW's engineers had beefed up the 2002's floorpan above the rear axle and had equipped the car with the larger ventilated front disc brakes from the 3.0 CS coupé and with a front-to-rear brake balancing valve. There were larger rear drum brakes, too. A limited-slip differential improved traction, and the engineering package was completed with an enlarged fuel tank and a taller final drive, without which the car would have accelerated even faster. Inside, the 2002 turbo was equipped with a red instrument surround, and its dials included a special 250kph (or 150mph) speedometer, an additional central pod for the turbo boost guage and a clock. Racing-style front seats completed the package.

The early 2002 from '71 to '76

The side decals left no-one in any doubt about the **2002 TURBO'S** identity

The early 2002 from '71 to '76

Right:
Note that the cars had the pre-1971 style of rear bumper, no doubt because the later type would have fouled the wheel arch extensions.

Far right:
This German-registered car shows the reverse script badging on the front air dam which was deleted on the series production cars

The early 2002 from '71 to '76

During the 2002 turbo's development, a top management shake-up at BMW had resulted in Sales Director Paul Hahnemann's departure, and in his place came Robert Lutz, formerly with Opel. It was Lutz who decided to give the turbocharged car its own image: instead of discreet badging to distinguish it from other 2002 models, the turbo would be dressed up as a roadgoing racer.

So it was that the car acquired wide wheel arches, a deep front spoiler and a boot lid spoiler, and the sort of racing stripes which were then popular on sporty American cars. Lutz decreed that the 2002 turbo should be available only in white or silver-grey, and that it should wear large "turbo" decals. Those on the front spoiler should be reversed, so that drivers being followed by a 2002 turbo would be able to read the name in their mirrors and would know they should move over to let the car past!

Almost as soon as the car was announced at Frankfurt, the trouble began. Its whole appearance was altogether too aggressive for the German road safety lobby, and one German parliamentarian demanded that the Government should investigate whether cars designed so overtly for speed and wearing what he called "war-paint" should actually be allowed on the roads. One result of this was that the reverse-script 'turbo' decals on the front spoiler were deleted before series production began in January 1974.

The **2002 TURBO** was something of a legend, although it was rather less driveble than BMW would probably have wished.

By this stage, however, the fuel shortages and fuel price increases consequent upon the 1973 October War between Israel and the Arab states had brought powerful cars into disrepute. No matter that the 2002 turbo was actually surprisingly frugal for such a fast car: its image as far as the public was concerned made it into a gas-guzzler. This did not help sales; and the fact that the car was not easy to drive and suffered from some teething troubles in service did not endear it to the buying public.

In view of this reception, BMW did not bother to put the 2002 turbo through emissions tests so that they could sell it in the USA. Nor did they develop right-hand drive versions, even though right-hand drive Britain showed enough interest to take 51 cars with left-hand steering. (In fact, the position of the turbocharger would have made it almost impossible to engineer a right-hand drive conversion). After just 1,672 examples had been built, production was halted in December 1974.

The early 2002 from '71 to '76

US models (1973-1976)

The US-model 2002tii, which was putting out only 125bhp to SAE standards by the end of the 1973 model year, was dropped for the 1974 season which began in summer 1973. This left the US market with just two 2002 models, the manual 2002 saloon and the 2002 Automatic.

The US editions of the Model 73 cars were distinguished by the same styling changes as their European counterparts, but they also had large bumpers, designed to meet the latest Federal regulations about low-speed collisions. The BMW engineers had done their best with these bumpers, making them of aluminium in order to save weight, but they did nothing for the 2002's styling and added both 100 lb in weight and 9.5 inches to the car's overall length.

As the 3-series cars were not ready for introduction into the USA until autumn 1976, the 2002 remained in production to bridge the gap even though the 3-series had taken over in Europe and other markets. However, manual-transmission 2002s ceased production in February 1976, and the last 2002 Automatic was built in July. That car was the last BMW 2002 of all - although it was not the very last of the 02 cars because the economy-model 1502 remained in production until July 1977.

It is difficult to find anything positive to say about the extended bumpers fitted to the US-market **2002** from 1973.

AJ 1158

Chapter Four

Selling the 2002

Sales catalogues for the BMW 2002 usually promoted the car's remarkable combination of sporting qualities, compact size and first-rate engineering. On the pages which follow are a representative selection of spreads from 2002 sales catalogues.

Selling the 2002

The earliest models
from a German catalogue
dated February 1969

Das Autofahren hat sich gewandelt. Genauso wie sich die Automobiltechnik mit der technischen Entwicklung und die Verhältnisse auf unseren Straßen mit der Motorisierung geändert haben. Das Autofahren wird heute beherrscht von der sachlichen Überlegung, schnell und bequem zum Arbeitsplatz zu kommen und für Beruf und Familie eine möglichst große Bewegungsfreiheit zu erreichen. Das ist die erste sachliche Grundlage für die Arbeit des Automobilkonstrukteurs. Die anderen, weit wichtigeren Bedingungen diktiert der Verkehr.

Der Verkehr heute erfordert einen besonderen Typ Automobil. Park- und Straßenraum sind knapp. Rangieren und Parken bereiten oft Mühe. Auf Ausfall- und Fernstraßen sind die physischen und psychischen Belastungen, die Anforderungen an das Reaktions- und Handlungsvermögen des Fahrers gewachsen. Gefordert wird beim Automobilbau deshalb, daß die modernsten technischen Erkenntnisse genutzt werden. Nur das garantiert bestes Fahrverhalten und größte Fahrsicherheit. Ein Beispiel für diesen Automobiltyp: Der BMW 2002.

Der BMW 2002 kombiniert Technik auf der Höhe unserer Zeit mit europäischen Abmessungen bei Größe und Innenraum. Der BMW 2002 ist ein sachlicher, über jeden automodischen Effekt erhabener, zweitüriger Wagen , der die Merkmale schneller und spurtstarker Reiselimousinen mit den Eigenschaften des übersichtlichen, wendigen, handlichen Stadtwagens kombiniert.

An der sorgfältigen Konstruktion, an der Kombination leistungsfähigster Baugruppen und an der automobiltechnischen Feinabstimmung des BMW 2002 liegt das souveräne Fahrgefühl: der Grund für die Freude am Fahren. Und die Tatsache, daß man dem Autofahren heute das Risiko nehmen kann.

Selling the 2002

1 Fahrwerk
Straßenlage ist die wichtigste Eigenschaft eines modernen Automobils. Sie wird entscheidend von der Hinterachskonstruktion beeinflußt. Jeder BMW hat schräggestellte Längsschwingen hinten. Eine Konstruktion, die Schule zu machen beginnt.

2 Triebwerk (Kurbelwelle)
Der BMW 2002 hat eine schwingungsarme, sorgfältig ausgewuchtete und 5-fach gelagerte Kurbelwelle mit 8 Gegengewichten.

3 Crashtests
In Zusammenarbeit mit der technischen Universität Berlin wurden in aufwendigen künstlichen Zusammenfahrunfällen die Konstruktion der BMW-Karosserien getestet. Das Ergebnis: Die Insassenkabine eines BMW ist Sicherheitszelle, Front- und Heck-Partie sind nach modernsten Erkenntnissen der Stoßabsorption konstruiert.

werk-Konstruktion, durch ausgewogene, genau dosierte Federung und Dämpfung und durch Drehstabstabilisatoren vorn und hinten in jeder Fahrsituation sicher auf die Straße.
Es ist auch ein BMW, der mit dieser Fahrwerk-Konstruktion als einzige Limousine der Welt die 172 Kurven des Nürburgrings unter 10 Min. schafft: 9 Minuten, 58,5 Sekunden. Die gleiche Zeit, die vor wenigen Jahren noch Juan Manuel Fangio mit seinem legendären Formel 1 Rennwagen fahren mußte, um den Großen Preis von Deutschland zu gewinnen. Ein Beispiel, wie leistungsfähig moderne Automobile sein können.
Das Überholen gehört zum Autofahren. Die Bedingungen dazu diktiert der totale Verkehr. Fahrzeuge, die nicht überholen wollen oder können, sind zu 80% der Beginn von Autoschlangen und Verkehrsverstopfungen. Die Folge: lange Überholwege und Kolonnen mit gereizten und hoch beanspruchten Fahrern. Die Ergebnisse: kurzsichtige Fahrentscheidungen, wagemutige Überholmanöver, Auffahrunfälle. Deshalb gehört keine jugendliche Leichtfertigkeit dazu, ein schnelles, spurtstarkes Automobil zu fahren. Denn ein sicheres Automobil ist schnell und muß in jedem Geschwindigkeitsbereich Leistungsreserven haben.
Der BMW 2002 hat 100 PS und beschleunigt in 10,9 Sekunden auf 100. Wenn Sie mit dem BMW 2002 120 km/h fahren, brauchen Sie von 100 vorhandenen 40 PS. 60 PS bleiben in Reserve. Mit den 60 PS verkürzen Sie Überholmanöver um Sekunden, vermindern das Risiko, erhöhen die Fahrsicherheit, können 170 km/h fahren und sicher sein, daß Sie nie Verkehrshindernis werden.
Seit über 50 Jahren baut BMW Hochleistungsmotoren. BMW baute Triebwerke für die legendäre JU 52. Die Triebwerke für das erste Düsenflugzeug der Welt. BMW hat mit modernen technischen Erkenntnissen die in den 20-er und 30-er Jahren herrschenden Motorrad-Traditionen gebrochen und mit den BMW 328 fast jedes Sportwagenrennen für sich entschieden. BMW baut heute mit diesen Erfahrungen einen der modernsten Hubkolbenmotoren für Automobile. Überlegene Leistung ist bei BMW eine Folge überlegener Technik.
Ein Beispiel: nach aufwendiger technisch physikalischer Grundlagenarbeit und erfolgreichem Einsatz in Rennmotoren baut BMW seit 1964 in jedes Triebwerk den Kugelwirbelwannenbrennraum. Die Folge: Ungewöhnliche Ergebnisse der Gasführung und Verbrennung. Das Ergebnis: durch eine gute Füllung des Zylinders und optimale Vermischung des Gasgemenges erreicht das BMW Triebwerk bei tiefen und hohen Drehzahlen überlegene Durchzugskraft.
Park- und Straßenraum sind knapp geworden. Ein verkehrsgerechtes Automobil ist deshalb übersichtlich, handlich und wendig. Es bietet soviel Innenraum wie möglich und soviel Karosserie wie nötig. Die Abmessungen sind vernünftig, die Linienführung sachlich. Die Heckpfosten sind im Interesse guter Sicht nicht breit geschwungen, die Karosserie nicht ausgebuchtet. Die Gürtellinie liegt tief. Die Sicht nach allen Seiten ist ausgezeichnet.
Die Karosserie ist Fahrgastraum und Sicherheitszelle. Sie muß so konstruiert sein, daß die Stoßenergie bei möglichen Unfällen in den „Knautschzonen" der Front- und Heckpartie aufgefangen wird und die Sicherheitszelle der Kabine unangetastet bleibt. BMW Automobile sind so konstruiert.
In Zusammenarbeit mit der Technischen Universität Berlin hat BMW in aufwendigen Versuchsreihen die selbsttragenden Karosserien getestet. Ziel der Versuche war, in Überschlägen, Front-Heck-, Front-Seite- und Front-Front-Zusammenstößen, die Konstruktion und die Belastbarkeit der Karosserien zu prüfen. Wie vorausberechnet sind jeweils nur Koffer- und Motorräume zusammengefaltet worden, der Insassenraum blieb immer unversehrt. Die Türen ließen sich öffnen, die Fenster herunterkurbeln.
BMW Karosserien werden in einem elektrophoretischen Tauchbad nach einem neuen Verfahren lakkiert. Alle unzugänglichen Winkel und Ecken der Karosserie, die in der normalen Lackierung unerreichbar bleiben, erreicht der korrosionsschützende Grundierlack. Die gesamte Karosserie wird fest und unlösbar mit Lack bedeckt. Die anschließend aufgebrachte und eingebrannte Mehrschichtlackierung widersteht allen Einflüssen von Wasser und Streusalz.

4 Triebwerk (Brennraum)
Das BMW-Triebwerk schöpft die Leistungsfähigkeit aus modernen physikalisch technischen Erkenntnissen. Ein Beispiel: Der Kugelwirbelwannenbrennraum.

5 Triebwerk (Formelmotor)
Überlegene Leistung ist bei BMW die Folge überlegener Technik, nicht die Folge eines überzüchteten Triebwerks. Mit geringen Mitteln kann z. B. aus dem 100 PS Serien-Triebwerk des BMW 2000 ein 260 PS Rennmotor gebaut werden.

6 Elektrophorese
In der Elektrophorese-Anlage erreicht der korrosionsschützende Grundierlack jede Ecke und jeden Winkel.

Sicherheit ist Sicherheit des Fahrverhaltens: Richtungsstabilität, gute Lenkeigenschaften, hervorragendes Kurven- und Bremsverhalten des Autos können die Fehler anderer Verkehrsteilnehmer oder eigene Unzulänglichkeiten ausbügeln. Eine zweckvolle Innenausstattung gibt zusätzliche Sicherheit. Sie ermöglicht, besser zu reagieren, nervenschonender, ermüdungsfreier zu fahren, die Möglichkeiten des Fahrzeuges günstiger zu nutzen und zum richtigen Zeitpunkt richtig zu reagieren. Vom Konzept her und von jedem Detail muß die Ausstattung so gestaltet sein, daß sie das Fahren erleichtert und damit sicherer macht.
Hohe Sitz-Positionen und tiefgezogene Verglasung ergeben hervorragende Rundumsicht und ermöglichen das korrekte Einschätzen der Wagenbegrenzung. Individuell verstellbare, körpergerechte Einzelsitze vermitteln direkten Kontakt mit der Straße.
Für die Reise ist viel und bequem erreichbarer Ablageplatz vorgesehen: eine offene Mulde auf dem Armaturenträger, ein großer, sich nach unten öffnender Direktsicht-Handschuhkasten, eine große Tunnelkonsole. Der BMW 2002 besitzt eine

intensive Heizung und Belüftung, die fein dosierbar ist und durch ein Gebläse unterstützt werden kann.
Auf dem Getriebetunnel sitzt griffgerecht der kurze, kräftige Schaltknüppel. Er erlaubt präzisen, leichten und schnellen Gangwechsel und vermittelt ein direktes Schaltgefühl. Auch das dient der Sicherheit.

Auf Wunsch wird der BMW 2002 mit Automatic geliefert. Automatisch läuft das Triebwerk immer im Tourenbereich mit dem günstigsten Drehmoment.

Im Blickfeld des Fahrers ist ein hochgezogener Instrumententräger mit übersichtlichen und blendfrei abgedeckten Instrumenten angeordnet. Alle Bedienungselemente liegen in Griffnähe. Um während der Fahrt die Aufmerksamkeit nicht dem Verkehr zu entziehen, erreicht der Fahrer Fern- und Abblendlicht, Lichthupe und Fahrtrichtungsanzeiger und eine Scheibenwisch- und Wasch-Automatic mit Zeigefingerkontakten. Beide Hände bleiben am Lenkrad.
Der BMW 2002 ist ein Automobil, mit dem Sie vom ersten Fahrkilometer an so vertraut sind, als hätten Sie nie einen anderen Wagen gefahren.

Die gleiche Sorgfalt wie bei unseren Konstrukteuren finden Sie auch bei unserem Kundendienst und Werkstattpersonal. Das weiß-blaue Zeichen ist eine Verpflichtung.

Selling the 2002

Model 71
from a UK catalogue

Outstanding engines

BMW got its name by making engines. And it is these engines that have made this name what it is today. Motoring journals have realized this. That's what makes them say that the BMW four-cylinder engines are just as powerful, smooth, silent, and flexible as normally only a six-cylinder can be.
The BMW four-cylinder engines have an overhead camshaft, inclined valves in an inverted V-arrangement, and a crankshaft with five bearings.
The combustion chambers feature the famous hemispherical swirl-action design.
The result is that today's BMW engines already fulfil tomorrow's pollution requirements.
Flexibility, acceleration, and speed are outstanding aspects typical of BMW engines. So that you always have top performance at any speed and overtaking becomes a matter of just a few seconds.

Selling the 2002

Traffic dimensions

Today's mass traffic sets a limit to the size of a car. A modern car must be maneuverable and suitable for congested traffic conditions.
There is something very sober and matter-of-fact about the styling of all BMW models. They all have a low waistline and visibility is excellent in all directions.
We tested the safety-first function of the BMW bodywork together with the Technical University in Berlin. A great number of crash tests were carried out, and they all had the same result. The front and the rear sections of the car were crushed and absorbed the whole impact, while the passenger compartment remained completely undistorted. It is a genuine safety compartment.
BMW applies the most modern methods in counteracting rust and corrosion. The bodywork is covered with a rustproof prime coating in a special electro-phoretic bath, and the multi-layer paint baked in on top of the primer coat guarantees that the bodywork cannot be affected by any weather conditions. A further contribution to this feature is made by the protective coating applied to the underfloor of the car. BMW uses a new method for protecting cavities, where there is a special risk of corrosion: the BMW cavity sealing procedure. This prevents rust from forming before it has even started.

The technical co-pilot

Rush-hour traffic, jammed roads and driving bumper-to-bumper take everything a driver can give. This stress reaches the physical and mental limit of what a human being can stand. So naturally automatic transmission is a very big contribution to making traffic safer. There is no need to operate the clutch or shift gears. And BMW's characteristic sportiness is maintained in full.
An automatic transmission makes things a lot easier for the driver. It enables him to concentrate completely on the traffic conditions surrounding him.

Selling the 2002

Four Headlamp Conversion

Rally Lamp

Sports Mirror

Alloy Road Wheel

Reclining Seats

BMW 1602
2002–2002 Tii
2000 Touring

	Part No.
Alloy Road Wheels 5½J Set of 5	90.00.0.000.106
Anti-roll Bar Kit, Front & Rear, *1602 only*	90.00.0.000.351
Badge Bar with Brackets for two lamps	90.00.0.000.221
Fog Lamp Round	90.00.0.000.331
Rear Fog Lamp Kit	90.00.0.000.338
Spot Lamp Round	90.00.0.000.334
Rally Lamp Spot	90.00.0.000.333
Rally Lamp Fog	90.00.0.000.336
2 Headlamp Halogen Conversion	90.00.0.000.341
4 Headlamp Halogen Conversion	90.00.0.000.340
Illuminated Switch	90.00.0.000.337
Laminated Windscreen *ex. Touring*	90.00.0.000.501
Laminated Windscreen *Touring only*	90.00.0.000.509
Tinted Laminated Windscreen *ex. Touring*	90.00.0.000 504
Inflatable Temporary Windscreen Kit	90.00.0.009.909
Heated Rear Window *ex. Touring*	90.00.0.000.601
Tow Bar Kit (Incl. Wiring Kit) *ex. Touring*	90.00.0.000.111
Tow Bar Kit (Incl. Wiring Kit)	90.60.1.109.424
Sports Horns	90.00.0.000.131
Roof Rack	90.00.0.000.191
Locking Petrol Cap	90.00.0.000.701
Mud Flaps *ex. Touring*	90.00.0.000.801
Mud Flaps *Touring only*	90.00.0.000.805
Sports Mirror Chrome	90.00.0.099.011
Sports Mirror Matt Black	90.00.0.099.012
Sheepskin Type Seat and Head Rest Cover	90.00.0.000.124
Steering Wheel Leather	90.00.0.000.201

We reserve the right to alter prices, specifications and designs without prior notice.

Continued on page 2

BLUE RIBBON

PARTS & ACCESSORIES

BMW CONCESSIONAIRES G.B. LTD

1

Selling the 2002

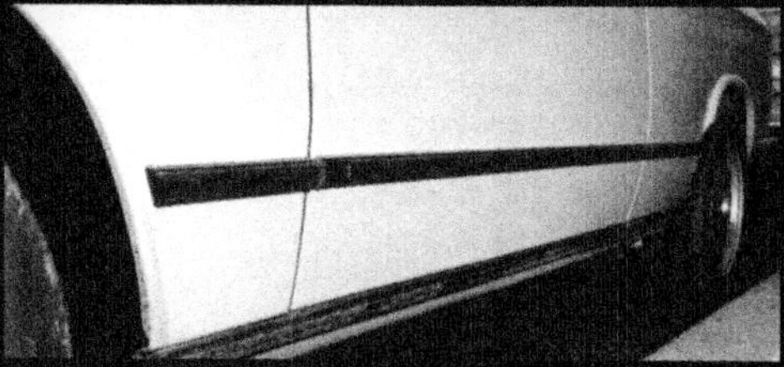

Protective Side Moulding

Rear Wing Stone Guard

Door Sill Protective Moulding

Speed Shift Gear Knob

Leather Steering Wheel

Luggage Compartment Carpet

BMW 1602 2002–2002 Tii 2000 Touring (cont.)

	Part No.
Gear Knob with BMW Emblem Leather (Manual only)	90.00.0.000.301
Gear Knob with BMW Emblem Wood (Manual only)	90.00.0.000.302
Speed Shift Leather Gear Knob (Manual only)	90.00.0.000.303
Emergency Kit (specify year)	90.00.0.000.661
Rubber Mats (Black) Set of 4 *ex. Touring*	90.00.0.000.881
Rubber Mats (Black) Set of 4 *Touring only*	90.00.0.000.885
Fully Reclining Sports Seat with Head Restraint. Complete with sub-frame (Left hand side)	90.00.0.000.129
Fully Reclining Sports Seat with Head Restraint, complete with sub-frame (Right hand side)	90.00.0.000.128
Luggage Compartment Carpet *ex. Touring*	90.00.0.000.884
Protective Side Mouldings (complete set) *ex. Touring*	90.13.9.898.589
Door Sill Protective Moulding (pair)	90.47.9.555.225
Rear Wing Stone Guard (Left)	90.60.1.804.501
(Right)	90.60.1.804.502

We reserve the right to alter prices, specifications and designs without prior notice.

BLUE RIBBON

PARTS & ACCESSORIES

BMW CONCESSIONAIRES G.B. LTD

2

Selling the 2002

BMW Touring

Touring
from a UK catalogue
dated June 1971

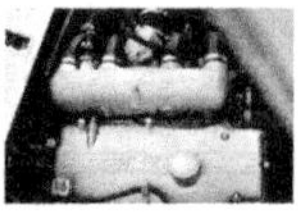

The engine

BMW engines all have an overhead camshaft, inclined valves in an inverted V-arrangement, and a crankshaft with five bearings. Swirl-action combustion chambers contribute to this high degree of perfection. And the result is that today's BMW engines already fulfil tomorrow's pollution laws.

The decisive characteristic of this design is not just supreme performance. But also flexible engines and spare power whenever you need it. So that you have full acceleration at any speed and overtaking becomes a matter of just a few seconds. How about this as an example: The BMW Touring 2000 tii rockets up from 0 — 60 mph in 9.2 seconds. It's really not surprising to hear what motoring journalists think: BMW four-cylinder engines have all the power, smoothness, and superiority that you'll normally only find in a six cylinder.

Selling the 2002

Functionability

The foremost aspect of all BMW designs is functionability. This starts with the engine, chassis, and braking system and goes on to the bodywork and the interior fittings. We think that luxury should also be functionable.
The seats can be adjusted individually and hold the body snugly. The backrests of the rear seats can be pulled forward separately and fastened in this horizontal position. So that you have even more space whenever you need it. The cushioned instrument panel features circular dials with an anti-dazzle effect. The switch for high beam and low beam, the flasher, the direction indicators, and the automatic screen washer have been designed for fingertip operation, so that the driver is not distracted and can keep both hands on the wheel.

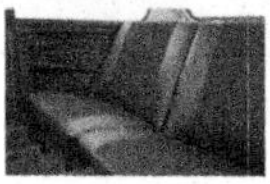

The interior
The BMW Touring features the largest interior of all comparable BMW models. It has a lot of room even for five persons. The backs of the rear seats can be folded down separately and fastened in this horizontal position, so that the additional space this gives you can be used in many ways and for many purposes.
By means of gas pressure springs the weight of the rear door is balanced exactly, so that the door can be moved with ease and always stays in the desired position. All the versions of the BMW Touring are equipped with a demisting unit for the widescreen and the side windows as standard and an electrically-heated rear window is also included in the basic equipment.

Selling the 2002

Targa-top cabriolet from a UK catalogue dated February 1972

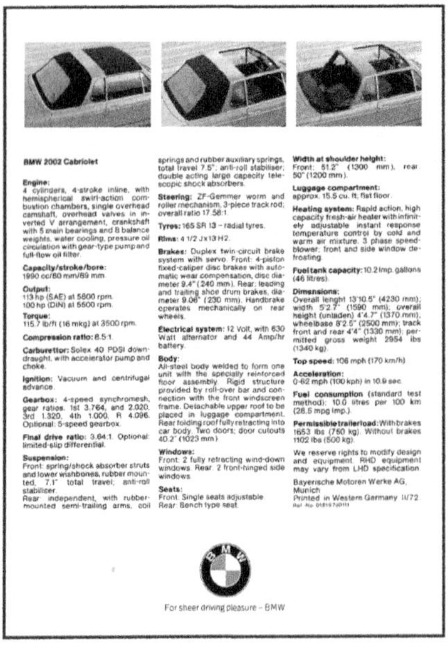

BMW 2002 Cabriolet

Engine:
4 cylinders, 4-stroke inline, with hemispherical swirl-action combustion chambers, single overhead camshaft, overhead valves in inverted V arrangement, crankshaft with 5 main bearings and 8 balance weights, water cooling, pressure oil circulation with gear-type pump and full-flow oil filter.

Capacity/stroke/bore:
1990 cc/80 mm/89 mm.

Output:
113 hp (SAE) at 5800 rpm.
100 hp (DIN) at 5500 rpm.

Torque:
115.7 lb/ft (16 mkg) at 3500 rpm.

Compression ratio: 8.5:1.

Carburettor: Solex 40 PDSI downdraught, with accelerator pump and choke.

Ignition: Vacuum and centrifugal advance.

Gearbox: 4-speed synchromesh, gear ratios. 1st 3.764, and 2.020, 3rd 1.320, 4th 1.000, R 4.096. Optional: 5-speed gearbox.

Final drive ratio: 3.64:1. Optional: limited-slip differential.

Suspension:
Front: spring/shock absorber struts and lower wishbones, rubber mounted, 7.1" total travel; anti-roll stabilicer.
Rear: independent, with rubber-mounted semi-trailing arms, coil springs and rubber auxiliary springs, total travel 7.5"; anti-roll stabiliser; double acting large capacity telescopic shock absorbers.

Steering: ZF-Gemmer worm and roller mechanism, 3-piece track rod; overall ratio 17.58:1.

Tyres: 165 SR 13 – radial tyres.

Rims: 4 1/2 J x 13 H2.

Brakes: Duplex twin-circuit brake system with servo. Front: 4-piston fixed-caliper disc brakes with automatic wear compensation, disc diameter 9.4" (240 mm). Rear: leading and trailing shoe drum brakes, diameter 9.06" (230 mm). Handbrake operates mechanically on rear wheels.

Electrical system: 12 Volt, with 630 Watt alternator and 44 Amp/hr battery.

Body:
All-steel body welded to form one unit with the specially reinforced floor assembly. Rigid structure provided by roll-over bar and connection with the front windscreen frame. Detachable upper roof to be placed in luggage compartment. Rear folding roof fully retracting into car body. Two doors; door cutouts 40.2" (1023 mm).

Windows:
Front: 2 fully retracting wind-down windows. Rear: 2 front-hinged side windows.

Seats:
Front: Single seats adjustable.
Rear: Bench type seat.

Width at shoulder height:
Front: 51.2" (1300 mm), rear 50" (1200 mm).

Luggage compartment:
approx. 15.5 cu. ft; flat floor.

Heating system: Rapid action, high capacity fresh-air heater with infinitely adjustable instant response temperature control by cold and warm air mixture. 3 phase speed-blower; front and side window defrosting.

Fuel tank capacity: 10.2 Imp. gallons (46 litres).

Dimensions:
Overall lenght 13'10.5" (4230 mm); width 5'2.7" (1590 mm); overall height (unladen) 4'4.7" (1370 mm); wheelbase 8'2.5" (2500 mm); track front and rear 4'4" (1330 mm); permitted gross weight 2954 lbs (1340 kg).

Top speed: 106 mph (170 km/h).

Acceleration:
0-62 mph (100 kph) in 10.9 sec.

Fuel consumption (standard test method): 10.0 litres per 100 km (28.5 mpg imp.).

Permissible trailer load: With brakes 1653 lbs (750 kg). Without brakes 1102 lbs (500 kg).

We reserve rights to modify design and equipment. RHD equipment may vary from LHD specification.

Bayerische Motoren Werke AG, Munich
Printed in Western Germany II/72

For sheer driving pleasure – BMW

Selling the 2002

Cabriolet with roll-over bar. A combination of superior safety and open air pleasure on the basis of functional sportiness.

Main part of this basis is the new BMW safety-first chassis. A chassis which BMW drivers refer to as a technical type of life insurance. The twin-dual-circuit brakes guarantee a braking effect twice as high as required by traffic regulations even if one of the circuits fails.

The engine in the BMW 2002 Cabriolet delivers a powerful 100 DIN hp and accelerates the car from 0 to 62,5 mph in 10,9 seconds.

The body styling combines the functional and matter-of-fact looks of the BMW 2002 with the merits of a well-designed roadster. It takes just a few seconds to remove the plastic roof and stow it in the luggage compartment. The rear window folds down neatly behind the rear seats without taking up any space. Then you have open-air driving at its best, with a strong and sturdy roll-over bar providing complete safety for the passengers in any situation. In fact, the BMW roll-over bar is an important additional safety factor in the convertible car concept which many drivers of open cars now regard as essential.

The interior is made to match the sporting appeal of the BMW Cabriolet. Body-contoured seats hold you securely in fast bends and make you forget about fatigue behind the wheel.

To summarize, this is genuine sportiness based on a carefully designed car concept. Sheer driving pleasure the purest way it comes.

Selling the 2002

Model 73
from a UK catalogue
dated August 1974

Selling the 2002

The chassis matched to the high performance of the power unit, offers more safety than BMW drivers require for everyday motoring. Its design reflects a section of automobile history. When this started to become generally accepted, BMW already had ten years head start in producing from their excellent chassis. This brilliant chassis which still sets an example today: Independent suspension. Front: Spring struts and coil springs. Rear: Semi-trailing arms. (BMW 2002 and more powerful models have torsion bar stabilisers front and rear)

The pivots of the semi-trailing arms in the rear are positioned inclined towards the body: ensuring accurate lateral adhesion.

The entirety reflects the care given to the details. BMW chassis offer the driver the safety to react calmly even on difficult cornering. They guarantee maximum road-holding and sensitive contact with the road. The steering geometry of the spring strut axle reduces the sensitivity of the chassis to damage from bad road surfaces, and neutralises developing centrifugal forces, on even the sharpest corners.

The power of the brakes are ideally matched to the power of the car. Dual circuit braking system with discs at the front and drums at the rear.

The brake servo makes braking noticably easier.

This elaborately detailed input, and the fine tuning of one group of elements to another, explains the exemplary steering, braking, and driving characteristics, in every situation, on every road, at any speed.

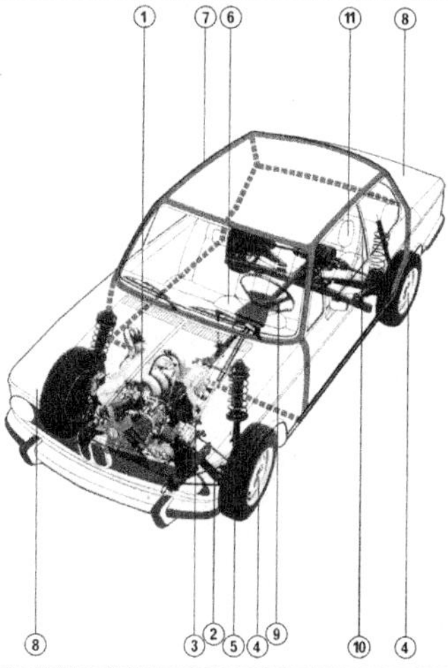

Active safety

1 Power Unit
The powerful power unit with its inherently smooth running conserves the fitness of the driver and reduces overtaking manoeuvres and accident risk.
2 Suspension
The BMW safety chassis has become an example. Even in critical situations it allows for the correction of driving errors.
3 Steering
The exact steering can cope with any driving situation. Safe, un-erring, straight running, even when obstructed.
4 Brakes
The servo-assisted dual circuit braking system with discs on the front and drums at the rear meet all braking requirements.
5 Tyres
Tyres, chassis, and power unit are balanced with the utmost care for optimum road-holding and highest comfort.
6 Layout of the Cockpit
BMW lays out the interior according to the strictest lines of functional design. Operating safety, spontaneously legible instruments, safe piloting thanks to the body-contoured seating.
All these details are the result of year long research and together add up to maximum driving safety.

Passive safety

7 Safety Cell
The passenger cell is specially rigid and remains almost undamaged in any kind of accident. The doors remain closed on impact. However, they can be opened after an accident.
8 Crumple Zones
Front and rear form the crumple zone which absorb the force of the impact. Steering column and steering gear lie outside the deformation zones. Through this, the forward thrust of the steering column is reduced to a minimum.
9 Steering System
The boss of the 4 spoke safety steering wheel deforms at minor impact.
10 Safety Belts
As standard, BMW fits 3point anchorage safety belts in the front.
11 Headrests
Heigh adjustable headrests in the front, fitted as standard protect the neck.

Active safety

1 Power Unit
The powerful power unit with its inherently smooth running conserves the fitness of the driver and reduces overtaking manoeuvres and accident risk.
2 Suspension
The BMW safety chassis has become an example. Even in critical situations it allows for the correction of driving errors.
3 Steering
The exact steering can cope with any driving situation. Safe, un-erring, straight running, even when obstructed.
4 Brakes
The servo-assisted dual circuit braking system with discs on the front and drums at the rear meet all braking requirements.
5 Tyres
Tyres, chassis, and power unit are balanced with the utmost care for optimum road-holding and highest comfort.
6 Layout of the Cockpit
BMW lays out the interior according to the strictest lines of functional design. Operating safety, spontaneously legible instruments, safe piloting thanks to the body-contoured seating.
All these details are the result of year long research and together add up to maximum driving safety.

Passive safety

7 Safety Cell
The passenger cell is specially rigid and remains almost undamaged in any kind of accident. The doors remain closed on impact. However, they can be opened after an accident.
8 Crumple Zones
Front and rear form the crumple zone which absorb the force of the impact. Steering column and steering gear lie outside the deformation zones. Through this, the forward thrust of the steering column is reduced to a minimum.
9 Steering System
The boss of the 4 spoke safety steering wheel deforms at minor impact.
10 Safety Belts
As standard, BMW fits 3point anchorage safety belts in the front.
11 Headrests
Heigh adjustable headrests in the front, fitted as standard protect the neck.

Selling the 2002

US Spec
dated 1976

Selling the 2002

The cockpit, designed for total control.

Inside the BMW 2002, you will find no replicas of Edwardian crests.

The design of the cockpit of the 2002 is the end result of extensive biomechanical simulation testing. All controls are within easy reach, precisely where you'd want them. All instruments are instantly readable and grouped in such a way as to preclude the need for unnecessary movement or a lapse of concentration.

The controls, grouped in the manner of an airplane cockpit.

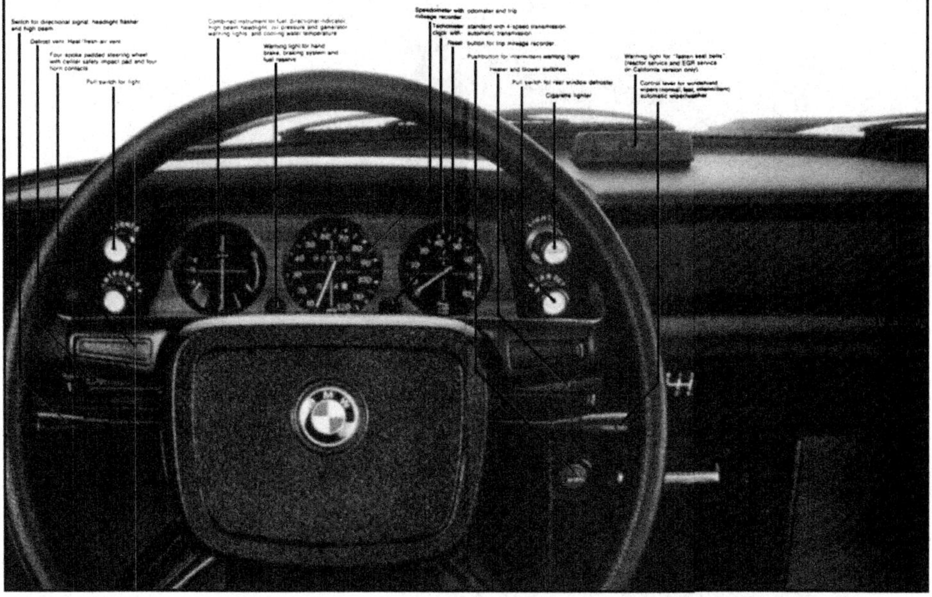

Safety, more than just brute strength.

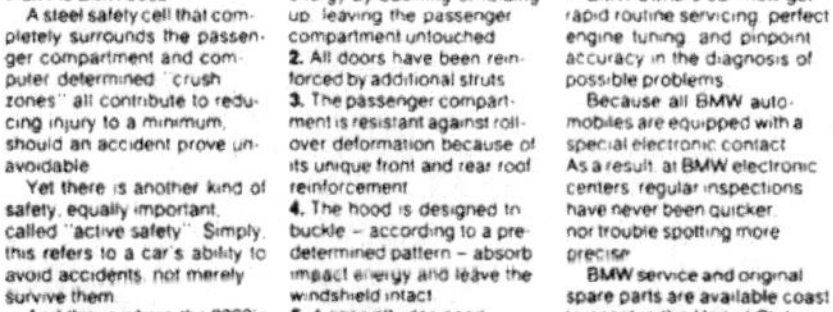

It's doubtful that there is a car made that's made stronger than the BMW 2002.

A steel safety cell that completely surrounds the passenger compartment and computer determined "crush zones" all contribute to reducing injury to a minimum, should an accident prove unavoidable.

Yet there is another kind of safety, equally important, called "active safety". Simply, this refers to a car's ability to avoid accidents, not merely survive them.

And this is where the 2002's extraordinary performance and handling characteristics come into play.

As does its dual twin-circuit, disc/drum braking system that provides adequate braking power – actually above the legally prescribed limit – even if one of the two systems should fail.

The BMW life saving system.

1. The BMW 2002 has deformable, energy absorbing front and rear sections. These sections are designed – by computer – to absorb impact energy by buckling or folding up, leaving the passenger compartment untouched.

2. All doors have been reinforced by additional struts.

3. The passenger compartment is resistant against rollover deformation because of its unique front and rear roof reinforcement.

4. The hood is designed to buckle – according to a predetermined pattern – absorb impact energy and leave the windshield intact.

5. A specially designed cardan tunnel plus a rigid front partition will prevent the engine from being driven back into the passenger compartment upon frontal impact.

6. Safety locks keep the doors closed during impact, yet permit subsequent opening.

7. The bumpers are braced against sturdy hydraulic shock absorbers, eliminating the possibility of damage to the car in frontal collisions of up to five miles per hour.

Service as efficient and reliable as the car itself.

BMW owners can now get rapid routine servicing, perfect engine tuning, and pinpoint accuracy in the diagnosis of possible problems.

Because all BMW automobiles are equipped with a special electronic contact. As a result, at BMW electronic centers, regular inspections have never been quicker, nor trouble spotting more precise.

BMW service and original spare parts are available coast to coast in the United States – and in over 100 countries around the world.

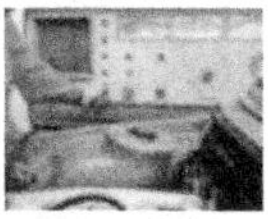

Technical Data BMW 2002

Dimensions and Weights	All steel shell welded with lower floor panel assembly, 2-door sedan. Length: 176.0" Width: 62.6" Height: (unloaded) 55.5" Wheelbase: 98.4" Track front and rear: 52.8" Turning circle dia.: 34.9' Door cutouts: 40.2" wide. Width at shoulder height: front 50.8", rear 49.6". Trunk capacity: approx. 15.9 cu. ft. Fuel tank capacity: approx. 13.0 US/gal., including 1.6 US/gal. reserve.	GVWR 3300 lbs. GAWR front 1680 lbs. rear 1780 lbs. Service load 790 lbs.
Engine, Power Transmission, Performance	Four-cylinder, four-stroke in-line engine, triple-hemispherical combustion chambers (transverse flow principle), overhead camshaft, parallel-displaced inclined overhead valves in V-arrangement. Forced oil lubrication with Eaton-type pump and full-flow oil filter, water cooled. Crankshaft has 5 main bearings plus 8 balance weight. Capacity: 1990 c.c./121.3 cu. in. Stroke: 3.15". Bore: 3.50". Power: 98 h.p. – SAE net at 5,500 rpm. Torque: 106 ft/lbs at 3,500 rpm. Compression ratio 8.1:1. Carburetor: SOLEX 32/32 DIDTA, 2-stage downdraft carburetor with booster pump and automatic choke with manifold system air injection (Thermal-Reactor on California version only.)	Ignition distributor with centrifugal advance and vacuum retard system. 12 volts, 630 watts, three-phase current alternator, battery 12 volts, 55 amp hrs. Gearbox: a. Manual transmission 4-speed synchromesh I 3.764 II 2.022 III 1.320 IV 1.000 R 4.096. b. Automatic transmission 3-speed with torque converter (optional) I 2.56 II 1.52 III 1.00 R 2.00. Final drive ratio: 3.90:1 (hypoid gears). 3.64:1 (California). Maximum speed 102 m.p.h. (Automatic 99 m.p.h.) Acceleration from 0 to 60 m.p.h.: 12.8 seconds (4 speed manual transmission). Regular gasoline: 91 RON.
Chassis and Brakes	Front Suspension: McPherson struts and wishbones, coil springs and stabilizers. Rear Suspension: Individual wheel suspension with rubber mounted steering axles (transverse), helical springs plus additional rubber springing and stabilizers. ZF-Gemmer steering system with worm and roller, three-part track rod, overall ratio 17.57:1. Rims: 5 J x 13 H 2.	Steel belted tires: 165 SR 13. Dual twin-circuit braking system with servo unit. Front: 4-piston fixed-caliper disc brakes with automatic adjustment. Disc diameter: 9.43". Rear: Simplex leading and trailing shoe drums, hand-brake acting on rear wheels mechanically. Drum diameter: 9.06".
Equipment	Heating and ventilation: water controlled fresh air heating system with easily adjustable temperature setting, three-speed blower, forced air exhaust at the rear windowpane via trunk lid louvers (except with steel sun roof). Defroster nozzles for windshield and front windows. Impact-absorbing bumpers with rubber welts, braced by means of hydraulic shocks, lateral rubber welts protect against impact, spring-loaded hood and safety lock, crank-operated windows in front completely disappear inside door, vent windows in front and rear, laminated glass windshield, heated rear windowpane, tinted glass all around. Rust protection: undercoating. Instrument panel unit with speedometer, odometer and trip recorder, tachometer standard with 4 speed transmission, clock with automatic transmission, fuel gauge, coolant temperature gauge, cigarette lighter, ignition key simultaneously turns out headlights together with the ignition, 2 backup lights, interior lighting, automatic wiper/washer unit for windshield with control lever arranged at steering wheel column, 2 wiper speeds plus interval setting, infinitely variable instrument panel lighting.	Additional warning lights: fuel reserve, hand brake and braking system, warning lights for "Fasten Seat Belts" (Reactor Service and EGR Service on California version only). Easily accessible storage space: in the glove compartment, above the instrument panel, in middle console, safety ashtray integrated in instrument panel, 2 ashtrays in passenger compartment, carpeting. Reclining seats in front have infinite adjustment and setting, armrests on doors, on right side with integrated handhold, straps suspended from roof with garment hooks in rear, headrests in front, adjustable and detachable, 3-point safety belts in front, 4 spoke padded steering wheel with impact safety pad plus 4 horn contacts, outside rearview mirror, tinted inside mirror, anti-glare, cloth or leatherette upholstery.
Optional Equipment	Automatic transmission, limited slip differential, light alloy rims, metallic paint, leather upholstery, lockable glove compartment, lockable tank cap, steel sunroof (manually or electrically operated), a choice of radios.	

GVWR = gross vehicle weight rating
GAWR = gross axle weight rating

Sole U. S. Importer:
BMW of North America, Inc.
Montvale, N.J. 07645
12541 Beatrice Street
Los Angeles, California 90066

Alterations in models, standard and optional equipment, as described in the text and illustrations, may occur. Precise information should be obtained from your BMW distributor or importer.

THE ULTIMATE DRIVING MACHINE

Printed in West Germany

Selling the 2002

Turbo
dated 1974

Selling the 2002

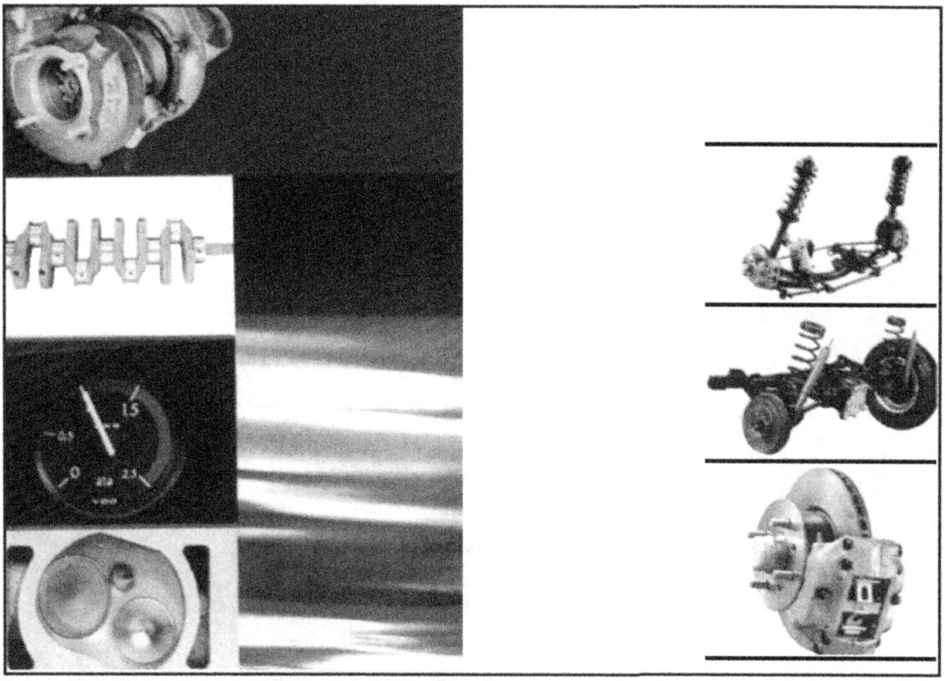

Selling the 2002

Alpina Catologue

Selling the 2002

special options

Any BMW-owner who seeks the ultimate in performance should seriously consider two simple bodywork conversions.

First, a switch to flared fibreglass wings – the reduced weight releases extra bhp for improved top speed and acceleration. Besides the fact that they cover wide wheels.

Second, fitting an aluminium or fibreglass front Air Dam. This will reduce aerodynamic drag by directing air flow around the sides of the car – rather than under it into an almost impenetrable area of turbulance.

Besides adding vital mph, an Air Dam gives more positive stability by progressively loading the front suspension with a downward thrust.

The BMW/ALPINA range of Special Options also includes oil coolers – invaluable for maintaining peak lubrication properties during sustained high performance.

SPORT PARTS

PRICE LIST
(all prices exclusive of fitting)

SECTION 1
Road Wheels

Model Application	Part No.	Basic Price	With V.A.T.
1602/2002/2002Tii/2000 Touring 5½J x 13 Perforated Steel Wheel, Silver Finish, Set of 5 Complete	BA1551351	45.50	
1602/2002/2002Tii/2000 Touring 5½J x 13 Forged Alloy Wheels (Borrani) Silver and Black, Set of 5 Complete	BA1551352	165.00	
1602/2002/2002Tii/2000 Touring 5½J x 13 Alloy Wheels Polished and Black, Set of 5 Complete	BA1551353	68.50	
2500/2800/3.0S/3.0Si 7J x 14 Alloy Wheel Silver and Black Finish, Set of 5 Complete	BA171454	249.50	

SECTION 2
Special Ventilated Disc Brake Conversions for extreme road use and competition events

Model Application	Part No.	Basic Price	With V.A.T.
2002Tii (Essential if engine output over 150bhp) Conversion Kit for Front Ventilated Disc Brakes includes Special Calipers, Discs, Back Plates and Pads	BA122121FDS	83.50	
2500/2800/2800CS (Essential if engine output over 200 bhp) Conversion Kit for Front Ventilated Disc Brakes includes Special Calipers, Discs, Back Plates and Pads	BA12532FDS	123.50	
3.0S/3.0CS (Thicker Ventilated Front Discs) Conversion to 28mm Front Discs with Modified Calipers.	BA12333FDS	132.75	
FRONT AND REAR VENTILATED DISC BRAKE CONVERSIONS			
1602/2002/2002Tii/2000 Touring Inc. Ventilated 20mm Front and Rear Discs, Calipers with handbrake mountings, trailing arms, drive shafts and pads	BA122122FRDS	410.96	
1602/2002 Extra Large Master Brake Cylinder	BA1221238MC		
FRONT AND REAR VENTILATED DISC BRAKE CONVERSIONS			
2500/2800/2800CS Discs, Calipers, Trailing Arms, Pads, etc.	BA1253124	432.00	

SECTION 3
Shock Absorbers

BILSTEIN SHOCK ABSORBERS AS USED BY ALL MAJOR RACE AND RALLY TEAMS

TYPE A FOR NORMAL FAST ROAD USE TO GIVE IMPROVED RIDE AND CORNERING POWER

Model Application	Part No.	Basic Price	With V.A.T.
1602/2002/2002Tii/2000 Touring 1 Pair Front Strut Inserts Road Setting	BA2321FRS	33.44	
1 Pair Rear Shock Absorbers Road Setting	BA2323RRS	20.20	
2500/2800/3.0S/3.0Si (Excluding Cars fitted with Nivomat) 1 Pair Front Strut Inserts Road Setting	BA23531FRS	49.48	
1 Pair Rear Shock Absorbers Road Setting	BA23532RRS	25.00	
2800/3.0CS/CSA/CSi/CSL 1 Pair Front Strut Inserts Road Setting	BA23533FRS	49.48	
1 Pair Rear Shock Absorbers Road Setting	BA23534RRS	25.00	

TYPE B THE ULTIMATE IN DAMPING FOR ROAD AND CLUB COMPETITION USE

Model Application	Part No.	Basic Price	With V.A.T.
1602/2002/2002Tii/2000 Touring 1 Pair Front Strut Inserts Sport Setting	BA2325FSS	38.52	
1 Pair Rear Shock Absorbers Sport Setting	BA2326RSS	20.20	
2500/2800/3.0S/3.0Si 1 Pair Front Strut Inserts Sport Setting	BA23537FSS	49.48	
1 Pair Rear Shock Absorbers Sport Setting	BA23538RSS	25.00	
2800CS/3.0CS/CSA/CSi/CSL 1 Pair Front Strut Inserts Sport Setting	BA22539FSS	49.48	
1 Pair Rear Shock Absorbers Sport Setting	BA225310RSS	25.00	

SECTION 4
Complete Chassis Conversion Sets

Model Application	Part No.	Basic Price	With V.A.T.
1602/2002/2000 Touring 5 Alpina perforated steel wheels 4½ x 13 with chrome centre caps and wheel nuts 2 Boge heavy duty front suspension units with negative chamber 2 Reinforced wishbones 2 Wheel hubs with bearings 2 Modified brake-calipers with heavy duty brake pads 2 Ventilated brake discs 20mm in thickness 2 Reinforced rear trailing arms 2 Bilstein heavy duty rear shock absorbers Large master brake cylinder	BA24212CCS	285.50	
2002Tii 5 Alpina perforated steel wheels 5½ x 13 with chrome centre caps and nuts 2 Boge heavy duty front suspension units 2 Ventilated brake discs 20mm in thickness Conversion set for brake calipers 2 Bilstein shock absorbers	BA242121CCS	195.00	
2500/2800 2 Bilstein heavy duty front suspension units 2 Bilstein heavy duty rear suspension units with adjustable spring support Adjustable anti sway bars 2 Ventilated brake discs 22mm in thickness Conversion set for brake calipers including special brake pads	BA2458123CCS	295.00	
3.0S/Si 2 Bilstein heavy duty front suspension units 2 Bilstein heavy duty rear suspension units with adjustable spring support Adjustable anti sway bars 2 Special front coil springs	BA2430124CCS	265.00	

SECTION 5
Adjustable Roll Bar Kits

Model Application	Part No.	Basic Price	With V.A.T.
1602/2002/2002Tii/2000 Touring Complete Front Adjustable Roll Bar Kit	BA25211FAR	26.85	
Complete Rear Adjustable Roll Bar Kit	BA25212RAR	20.50	
2500/2800/3.0S/3.0CS Complete Front Adjustable Roll Bar Kit	BA25513FAR	30.65	
Complete Rear Adjustable Roll Bar Kit	BA25514RAR	30.65	

SECTION 6
Carburettor Kits

CARBURETTOR KITS WITH EITHER WEBER OR DELLORTO TWIN CHOKE CARBURETTORS

Model Application	Part No.	Basic Price	With V.A.T.
1602/2002/2000 Touring Two Twin 40 Choke Carburettors Complete with Manifolds, Linkages, Air Cleaners, etc.	BA362240CK	110.00	
Two Twin 45 Choke Carburettors, Complete with Manifolds, Linkages, Air Cleaners, etc.	BA362245CK	120.00	
1 28/36 Down Draft Carburettor with Manifold, Air Cleaner, Linkage, etc.	BA362128CK	56.50	
2500/2800/3.0S/3.0CS Three Twin 40 Choke Carburettors, Complete with Manifolds, Air Cleaner, Linkage, etc.	BA365340CK	152.00	
Three Twin 45 Choke Carburettors, Complete with Manifolds, Air Cleaner, Linkage, etc.	BA365345CK	160.50	

SECTION 7
Exhaust Manifolds

Model Application	Part No.	Basic Price	With V.A.T.
1602/2002/2002Tii/2000 Touring 4 Branch Exhaust Manifold suitable for use with Standard Exhaust System	BA3T1411EM	42.00	
2500/2800/3.0S/CS 6 Branch Exhaust Manifold suitable for use with Standard Exhaust System	BA375312EM	85.00	

SECTION 8
Cylinder Heads

Model Application	Part No.	Basic Price	With V.A.T.
1602/2002/2002Tii/2000 Touring Gas Flowed Head complete with Valves, Springs and Standard Cam (EXCHANGE)	BA48211CH	55.00	
2500/2800/3.0S/3.0CS Gas Flowed Head, Complete with Valves, Springs and Standard Cam (EXCHANGE)	BA48512CH	79.50	

SECTION 9
Camshafts

FOR HIGHER ENGINE OUTPUTS IF USED IN CONJUNCTION WITH OTHER ENGINE MODIFICATIONS

Model Application	Part No.	Basic Price	With V.A.T.
1602/2002/2002Tii/2000 Touring 300° Camshaft for road use	BA492112RC	49.50	
324° Camshaft for competition use	BA492113SC	55.00	
2500/2800/3.0S/3.0CS 300° Camshaft for road use	BA493114RC	78.50	
324° Camshaft for competition use	BA493115SC	80.50	

SECTION 10
Complete Engine Kits include Cylinder Head, Carburettors, Camshaft, Exhaust Manifold etc.

Model Application	Part No.	Basic Price	With V.A.T.
2002/2000 Touring Alpina Cylinder Head with Twin Weber Carburettors, Alpina Camshaft, Exhaust Manifold, Complete (140hp)	BA410211CHK	421.43	

SECTION 11
Spoilers (Air Dams)

SPOILERS TO GIVE INCREASED DOWNTHRUST AND GREATER HIGH SPEED STABILITY

Model Application	Part No.	Basic Price	With V.A.T.
1602/2002/2002Tii/2000 Touring Front Spoiler Fibreglass	BA511221FPS	29.50	
Front Spoiler Aluminium (in Primer)	BA 511222FAS	25.50	
2500/2800/3.0S/Si/SA Front Spoiler Fibreglass	BA511321FPS	36.50	
Front Spoiler Aluminium (in Primer)	BA511322FAS	28.00	
2800CS/3.0CS/CSA/CSi/CSL Front Spoiler Fibreglass	BA511323FPS	36.50	
Front Spoiler Aluminium (in Primer)	BA511324FAS	28.00	

SECTION 12
Oil Coolers for road or competition use protect engine by maintaining oil at correct operating temperature

Model Application	Part No.	Basic Price	With V.A.T.
1602/2002/2000 Touring Oil Cooler Kit for Road Use	BA5122120CK	19.75	
Oil Cooler Kit for Competition Use	BA5122130CK	21.25	
2002Tii Oil Cooler Kit for Road Use	BA5122140CK	19.75	
Oil Cooler Kit for Competition Use	BA5122150CK	21.25	
2500/2800/3.0S/Si/CS/CSi/CSL Oil Cooler Kit for Road Use	BA5123160CK	21.75	
Oil Cooler Kit for Competition Use	BA5123170CK	23.25	

SECTION 13
Fibreglass Bodywork

EXTRA LIGHT INCORPORATING FLARED WINGS TO COVER WIDE WINGS

Model Application	Part No.	Basic Price	With V.A.T.
1602/2002/2002Tii Complete Set Front and Rear Wings	BA513111FW	145.50	

SECTION 14

Model Application	Part No.	Basic Price	With V.A.T.
Right Front Wing	BA513112FW	41.10	
Left Front Wing	BA513213FW	41.10	
Right Rear Wing	BA513214FW	31.65	
Left Rear Wing	BA513215FW	31.65	

SECTION 15
Steering Wheels

Model Application	Part No.	Basic Price	With V.A.T.
1602/2002/2002Tii/2000 Touring 14" Black Leather Steering Wheel. New "K" type safety design complete with boss, horn push, etc.	BA615214SW	13.95	
2500/2800/3.0/CS/CSi/CSL 15" Black Leather Steering Wheel. New "K" type safety design complete with boss, horn push, etc.	BA615315SW	13.95	

SECTION 16
Gear Knobs

Model Application	Part No.	Basic Price	With V.A.T.
All Models (Except Automatics) Rosewood Gear Shift Knob with inset "Alpina" Emblem	BA616321WGK	1.65	

SECTION 17
"Alpina" Self Adhesive Side Strips

Model Application	Part No.	Basic Price	With V.A.T.
BMW "ALPINA" side strips Black or Silver, Specify Colour	BA817111SS	2.50	

SECTION 18
BMW – "ALPINA" Sport-Parts Rally Jacket

Blue and White with either black fur lining or black nylon lining complete with matching hood, storm cuffs, inside pocket etc. Sport-Part and BMW badge supplied.

Model Application	Part No.	Basic Price	With V.A.T.
Adults (Male) Fur lined	BA181211XL	13.50	
Adults (Male) Fur Lined	BA181212L	13.50	
Adults (Male) Fur Lined	BA181213M	13.50	
Adults (Male) Fur Lined	BA181214S	13.50	
Adults (Male) Nylon Lined	BA181215XL	12.25	
Adults (Male) Nylon Lined	BA181216L	12.25	
Adults (Male) Nylon Lined	BA181217M	12.25	
Adults (Male) Nylon Lined	BA181218S	12.25	
Ladies 34" Nylon Lined	BA18121934	12.25	
Ladies 36" Nylon Lined	BA18122036	12.25	
Childs 28" Nylon Lined	BA18122128	10.90	
Childs 30" Nylon Lined	BA18122230	10.90	
Childs 32" Nylon Lined	BA18122232	10.90	

SECTION 19
Hella Rally Lamp Kit.

Model Application	Part No.	Basic Price	With V.A.T.
2 OI Long Range Rally Lamps complete with bulbs and cable	BA1925312RL	24.70	

Turn Your Hymnals to 2002—David E. Davis, Jr. Blows His Mind on the Latest from BMW

As I sit here, fresh from the elegant embrace of BMW's new 2002, it occurs to me that something between nine and ten million Americans are going to make a terrible mistake this year. Like dutiful little robots they will march out of their identical split-level boxes and buy the wrong kind of car. Fools, fools! Terrible, terrible, I say. Why are you blowing your money on this year's too-new-to-be-true facelift of the Continental/Countess Mara/Sprite/Sprint Status Symbol/Sting Ray/Sex Substitute/ Mainliner / Belair / Newport / Overkill / Electra / Eldorado / Javelin / Toad / GTO / GTA/GTB/GTS/GTX/Reality Blaster/ Variant/Park Lane/Park Ward/Ward-Heeler/XK-E/Dino/Dud car when you should be buying a BMW 2002, I ask.

Down at the club, Piggy Tremalion and Bucko Penoyer and all their twit friends buy shrieking little 2-seaters with rag tops and skinny wire wheels, unaware that somewhere, someday, some guy in a BMW 2002 is going to blow them off so bad that they'll henceforth leave every stoplight in second gear and never drive on a winding road again as long as they live.

In the suburbs, Biff Everykid and Kevin Acne and Marvin Sweatsock will press their fathers to buy HO Firebirds with tachometers mounted out near the horizon somewhere and enough power to light the city of Seattle, totally indifferent to the fact that they could fit more friends into a BMW in greater comfort and stop better and go around corners better and get about 29 times better gas mileage.

Mr. and Mrs. America will paste a "Support Your Local Police" sticker on the back bumper of their new T-Bird and run Old Glory up the radio antenna and never know that for about 2500 bucks less they could have gotten a car with more leg room, more head room, more luggage space, good brakes, decent tires, independent rear suspension, a glove box finished like the inside of an expensive overcoat and an ashtray that slides out like it was on the end of a butler's arm—not to mention a lot of other good stuff they didn't even know they could *get* on an automobile, like doors that fit and seats that don't make you tired when you sit in them.

So far as I'm concerned, to hell with all of 'em. If they're content to remain in the automotive dark, let them. I know about the BMW 2002, and I suspect enthusiasts will buy as many as those pink-cheeked Bavarians in their leather pants and mountain-climbing shoes would like to build and ship over here. Something between nine and ten million squares will miss out on this neat little 2-door sedan with all the *cojones* and *brio* and *elan* of cars twice its size and four times its price, but some ten thousand keen types will buy them in 1968, so the majority loses for once.

The 2002 is BMW's way of coping with the smog problem. They couldn't import their little 1600 TI, because their smog device won't work on its multi-carbureted engine. So they stuffed in the smooth, quiet 2-liter (single carburetor) engine from the larger 2000 sedan and—SHAZAM—instant winner!

To my way of thinking, the 2002 is one of modern civilization's all-time best ways to get somewhere sitting down. It grabs you. You sit in magnificently-adjustable seats with great, tall windows all around you. You are comfortable and you can see in every direction. You start it. Willing and un-lumpy is how it feels. No rough idle, no zappy noises to indicate that the task you propose might be anything more than child's play for all those 114 Bavarian superhorses.

Depress the clutch. Easy. Like there was no spring. *Snick*. First gear. Remove weight of left foot from clutch. Place weight of right foot on accelerator. The minute it starts moving, you know that Fangio and Moss and Tony Brooks and all those other

Chapter Five

"What the Press said"

The BMW 2002 attracted a great deal of press comment during the years it was in production. Road testers worldwide found its performance and handling exceptional - and discovered that the combination offered in later models was even better.

However, not everything about the 2002 models found favour. Passenger-compartment ventilation was never sufficient, rear seat space was always limited, and road-testers either failed to mention the car's appearance or described it with words such as "mild" and "functional". The 2002 turbo also brought forth some mixed reactions, despite its formidable performance.

These extracts from a wide variety of English-language road tests published between 1968 and 1975 give a good impression of the way the cars were viewed when they were new. The full road tests from which these quotations are taken, as well as a selection of others, can be found in Brooklands Books' BMW 2002 Gold Portfolio, 1968-1976. All quotations are reproduced here with the kind permission of their copyright holders.

The 2002's performance

"The engine is delightful...The combination of exceptional medium-speed torque with a car of light weight produces quite remarkable flexibility, and the top-gear performance is reminiscent of a V8."

2002, Autosport, 17th May 1968

"... the performance for most of the range, and economy, are better than the very fast Cortina Lotus."

2002, Motor, 8th June 1968

"What the Press said"

"You give away a little performance with the automatic, as you would expect ... (but) it is still plenty lively ... and goes about the job with a delightful eagerness."

2002 Automatic, Road and Track, November 1969

"... if you love to drive, why buy an Automatic BMW 2002? The reasons are mostly because the ZF box doesn't steal enough performance to bother any but track addicts."

2002 Automatic, World Car Guide, February 1970

"... the very small loss of performance with the ZF box may actually amount to an appreciable gain under typical road conditions."

2002 Automatic, Autosport, 30th July 1970

"At first glance, there might appear to be little advantage in the system over the multi-carburettor version of the same unit, for the gain in maximum power is not great. The point, however, is that the fuel injection engine gives more torque and it continues to give it over a wide band of revs. In practical terms, that means instant acceleration at any time, which is what fast road motoring is all about."

2002tii, Autosport, 4th November 1971

"The 2-litre BMW engine must be one of the best four-cylinder units ever built. It has none of the roughness at low speeds which plagues most of the larger fours and is outstandingly flexible."

2002tii, Autosport, 14th March 1974

"... one of the most impressive features of the car is its tractability and fussless behaviour in town. You don't expect to be able to trundle along at 15mph in top with an engine that produces 85bhp per litre, but do it you can.

Good as the Turbo is for out-and-out performance driving, it can be a real pain under more normal circumstances. Floor the throttle at

3500rpm to overtake a slower car and there's a good couple of seconds delay before the turbocharger gets the car really moving; by that time you're normally past the obstacle and the sudden increase in power can be an embarrassment. We'd prefer more mid-range power even if it meant a slight sacrifice in top-end performance."

2002 turbo, Motor, 25th May 1974

"... at just over 3,000rpm, you can hear a slight whistle; on up to 4,000 the whistle becomes more apparent, and then the needle in that little gauge on the right moves out of the white section and into the green ... All of a sudden, you seem to have found another hundred horsepower, and the little car simply hurtles along, the tachometer needle at 6,500, and you wonder what the hell happened. It's as if you'd been driving with the handbrake on and you'd suddenly remembered."

2002 turbo, Competition Car, June 1974

The 2002's handling

"Good as the roadholding is of the larger models, the 2002 feels altogether more crisp and responsive to handle, which is saying a great deal."

"The machine is beautifully balanced, having just enough understeer to give high-speed straight running, but when driven reasonably hard the characteristic is neutral. Ultimately, the rear end can be made to hang out a little on power, but it does not fully break away. A very fine sense of control is imparted and there always seems to be a reserve of roadholding to cope with the unexpected."

2002, Autosport, 17th May 1968

"...roadholding and general behaviour on all dry surfaces are matched by very few other saloons or mass production sports cars."

2002, Motor, 8th June 1968

"What the Press said"

"Cornering is fantastic. On a winding, tight, rugged mountain road the car has no peer in its class."
2002, Motor Trend, July 1968

"Hard driving inspires confidence and it is exceptionally difficult to get completely out of shape in this car. We never came close."
2002, Road Test, May 1971

"Though the Tii is such a satisfying everyday touring car, there is the feel of a competition car about it. Very well balanced, it flicks through corners without effort, and it shows its breeding when forced to change its line in the middle or to take an adverse camber in its stride. The independent rear end sticks down extremely well over broken surfaces, while the absence of wheelspin on leaving sharp corners is praiseworthy. Nevertheless, an oversteering condition can be induced by suitable provocation."
2002 tii, Autosport, 22nd July 1971

"At the limit, which is naturally more easily reached in the Tii, the limitations of the semi-trailing rear suspension start to show themselves as the back end jacks-up slightly and starts to drift sideways. Full-blooded oversteer is only encountered on very fast bends, because the more usual result is for the inside rear wheel to lift and spin, slowing the car."
2002tii, Autocar, 30th December 1971

"The changes made to the suspension make the handling extremely good under most circumstances. To all intents and purposes the car steers where you point it with no tricks such as vicious tuck-in if you lift-off. But push the car to its (admittedly high) limit of adhesion and it can become a little untidy particularly if the turbocharger cuts in half-way round a corner. On wet London streets the tail could step sideways very quickly too."
2002 turbo, Motor, 25th May 1974

"What the Press said"

The 2002 and its driver

"... the 2002 is one of modern civilisation's all-time best ways to get somewhere sitting down. It grabs you. You sit in magnificently-adjustable seats with great. tall windows all around you. You are comfortable and you can see in every direction."

2002, Car and Driver, April 1968

"On a car of this price, I would have expected the wiper arcs to be matched to r.h.d., but they're not and they leave a nasty blind spot above and to the right of my eye-line."

2002, Autocar, 12th December 1968

Chapter Six

The tuned 2002

The excellence of the basic 2002 soon made it a firm favourite with Germany's tuning specialists. Most of these had already offered improvements for the two-door 02 series cars, but the 2002 obviously gave them even more scope. Naturally, some companies made available off-the-shelf go-faster accessories, but these were not companies which have endured. Of much greater interest are the companies which offered a complete performance conversion service, adding better handling and braking to their engine power increases in order to maintain the balance of the converted vehicle.

In many cases, these companies converted cars for pure competition purposes, but this was to some extent a sideline: the real money was to be made from selling conversions which could be used on public roads. Today, 2002s converted by the the leading companies in the 1970s have acquired an almost legendary status, even though their actual performance has now been equalled by all too many everyday production saloons.

Alpina

The oldest and the best known of the companies which worked on the BMW 2002 was Alpina. The company had been started in 1963 by Burkard Bovensiepen, the son of a typewriter manufacturer, and the first high-performance conversions were carried out in a designated area of the Alpina typewriter factory. By 1965, Bovensiepen had decided to work exclusively on BMW cars, and in 1969 the company acquired its own premises at Buchloe in Bavaria.

It was also in 1969 that Alpina established its own racing team, and over the next few years this proved to be very successful. Although they came up against the works BMW racing saloons in the early 1970s, Alpina successfully maintained a good relationship with the factory.

The tuned 2002

One reason was the high quality of their conversions, and BMW agreed not only to honour their standard warranty on any Alpina-converted car but also to distribute Alpina products through their own dealerships. In due course, the relationship became even closer, until by the middle 1980s Alpina were accepted as the official BMW tuning specialists.

Alpina did not believe in bolt-on high-performance conversions: all their engine work involved a complete stripdown and rebuild to blueprinted standards. Only in this way were they able to ensure the

The Alpina graphics were popular in the first half of the 1970s.

reliability and durability which had earned them the support of BMW. All their conversion work also met the strict German TÜV regulations, although some conversions were marketed for export only for a time.

No records exist of the conversions which Alpina carried out on 2002s, and Alpina-converted cars did not bear any special identifying plate to authenticate their origins, although cars converted in Germany did have documentation from Alpina to prove their authenticity. This can present some problems today. Conversions were definitely carried out for many territories outside Germany, and in Britain the Alpina franchise was held by Crayford Engineering from 1970, until BMW (GB) took over in 1973 with its newly-established BMW Sports Parts division. Confusingly, there was also a UK arm of Alpina, known as Alpina Automotive Ltd and trading from Pershore in Worcestershire.

When the 2002 was announced in 1968, Alpina already had a successful series of conversions available for the 1600-engined cars, and it was probably some time before the first converted 2002s appeared. However, Alpina-modified 2002s did become available in the early 1970s, in several states of tune and with a wide variety of options. From 1975, it was possible to buy Alpina side stripes for the 2002, but these in themselves are no guarantee of the authenticity of a car claimed to have been modified by Alpina!

The tuned 2002

Max Hoffman, the US importer had three Alpina - tuned **2002s** built for the show circuit in 1972. This was one of them.

The tuned 2002

Engine specifications

Alpina offered six basic roadgoing engine specifications for the 2002. With the exception of the least powerful A1 engine, all Alpina engine conversions had to be accompanied by ventilated front disc brakes in order to meet TÜV regulations in Germany.

In addition, the company produced engines for Group 1 (production cars) and Group 2 saloon racing.

	ALPINA A1	ALPINA A2	ALPINA A2S
Capacity	1,990cc	1,990cc	1,990cc
Power	115bhp at 5,800rpm	150bhp at 6,600rpm	157.5bhp at 6,600rpm
Torque	125 lb.ft at 4,000rpm	137 lb.ft at 5,000rpm	144 lb.ft at 5,600rpm
Rev limit	6,500rpm	6,800rpm	7,000rpm
Compression	8.5:1	10:1	10:1
Camshaft	264° (standard)	300°	300°
Valve sizes	inlet 46mm, exhaust 38mm	inlet 46mm, exhaust 38mm	inlet 46mm, exhaust 39mm
Max valve lift	9mm	10mm	10mm
Pistons	standard	standard tii type	dome-top (although marketed by KS, these were actually made by Mahle)
Con rods	standard	standard	modified standard
Fuel system	two Solex 40 DDH carburettors	two Solex 45 DDH carburettors	two Solex 45 DDH carburettors
Ignition timing	standard	34° at 5,000rpm	32° at 5,000rpm
Spark plugs	W200 T30	W240 T2	W240 T2
Exhaust manifold	standard	modified standard	modified standard
Exhaust system	standard	standard	Alpina

The tuned 2002

ALPINA A3	ALPINA A4	ALPINA A4S
1,990cc	1,990cc	1,990cc
157.5bhp at 6,600rpm	160bhp at 5,700rpm	168bhp at 5,700rpm
144 lb.ft at 5,600rpm	137 lb.ft at 5,500rpm	146 lb.ft at 5,500rpm
7,000rpm	6,800rpm	7,000rpm
10:1	10:1	10:1
300°	300°	300°
inlet 46mm, exhaust 39mm	inlet 46mm, exhaust 39mm	inlet 46mm, 39mm
10mm	10mm	10mm
dome-top (although marketed by KS, these were actually made by Mahle)	standard tii type	dome-top (although marketed by KS, these were actually made by Mahle)
modified standard	standard	modified standard
two Weber 45 DCOE carburettors	Kugelfischer/ Alpina injection	Kugelfischer/ Alpina injection
32° at 5,000rpm	34° at 5,000rpm	32° at 5,000rpm
W240 T2	W240 T2	W240 T2
modified standard	modified standard	modified standard
Alpina	Alpina	Alpina

The German magazine Auto Motor und Sport tested a 160bhp Alpina 2002 and recorded a top speed of 129mph.

In Switzerland, Automobil-Revue tried a 173bhp conversion which gave a maximum speed of 127mph and reached 62mph in 7.5 seconds from a standing start.

The tuned 2002

Cylinder Head Modifications

The cylinder heads of all Alpina engines were modified to some extent. Few details of the modifications have ever been made available, but observation reveals the following common factors:

enlarged and polished inlet ports
enlarged and polished exhaust ports
hemispherical combustion chambers (A2S and A4S only)
larger inlet valve seats (A2S and A4S only)
larger exhaust valve seats (race and rally engines only)
matched combustion chamber volumes
polished combustion chambers

Engine and Transmission Parts

alloy sump
Alpina air filter box for carburetted cars
Alpina air filter box for fuel injected cars
balanced and polished connecting rods
close-ratio gearbox (Getrag 5-speed)
competition cylinder head gasket
electric cooling fan
exhaust manifold (4 into 2 into 1)
free-flow exhaust system (stainless steel)
high-pressure oil pump
larger capacity water radiator
limited-slip differential (75% or 40% lock, any ratio)
modified rear drive flanges (reinforced)
oil cooler kit for differential
oil cooler kit for engine
shortened propshaft for 5-speed gearbox
specially painted and badged cam cover
sports camshaft (300°, small and large bearings)
stronger rubber engine mountings
stronger rubber gearbox mountings
uprated clutch

Steering and Suspension Parts

adjustable front anti-roll bar
adjustable rear anti-roll bar
Alpina 5.5Jx13 alloy wheels
Alpina 5.5Jx13 steel wheels

The tuned 2002

Alpina 6.5Jx13 alloy wheels
Bilstein gas-filled front damper inserts
Bilstein gas-filled adjustable rear dampers
Boge heavy-duty front damper inserts
Borrani 6Jx13 alloy wheels
high-ratio steering box (12.8:1)
Minilite 6Jx13 alloy wheels
modified front hubs (reinforced)
modified rear trailing arms (reinforced)
negative camber front struts (1°)
shorter, stronger front road springs
shorter, stronger rear road springs
stronger rubber bushes for rear subframe mounts
stronger rubber bushes for standard anti-roll bars
stronger rubber bushes for track control arms and tie-rod
stronger rubber bushes for trailing arms
strut top castor angle adjusting plates

Among Alpina's accessories were steel wheels like this one.

Braking system
larger brake master cylinder
modified front brake calipers (to accept thicker discs)
modified front brake disc dust covers (to aid cooling)
modified rear brake drums, shoes and cylinders (larger)
special brake pads (Textar or Ferodo)
ventilated front brake discs (255mm x 20mm)

Bodywork, interior and accessories
additional instruments
additional lighting (front spotlamps)
Alpina badged gearshift knob
Alpina badged sports steering wheel
Alpina body stripe set (various colours)
bucket type front seats
front and rear wheelarch extensions
front spoiler (glass fibre or alloy)
half-bumpers for the front
modified speedometer (220kph, white insert)
modified tachometer (8,000rpm, white insert)
rallying tripmeter
sports door mirrors

The tuned 2002

Group 1 Racing Engines

The engines for Group 1 racing saloons had to remain in standard specification, although they were allowed to use any parts specially homologated for racing (i.e. approved by the racing authorities). In preparing the engines, it was also permissible to take advantage of any manufacturing tolerances to improve performance.

The precise specification of the Alpina Group 1 racing engine was never released. However, with a homologated exhaust system, the Group 1 2002tii engine produced 140bhp, an increase of 10bhp over the standard 2002tii.

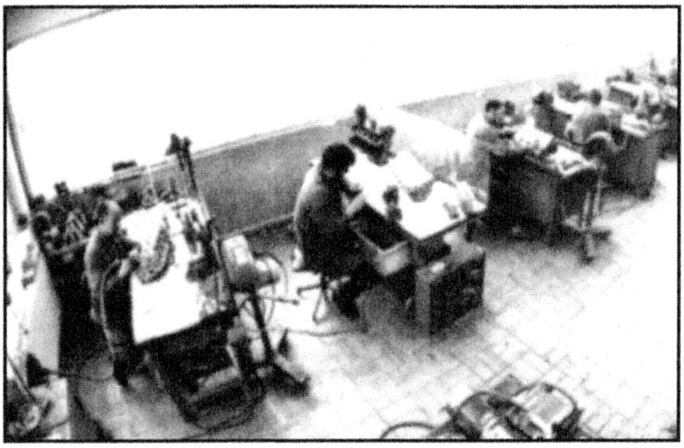

Group 2 Racing Engines

Once again the precise specification of the Alpina Group 2 engines is not available. However, the 1990cc engine produced 205bhp in race tune and 195bhp in rally tune. This was achieved by using forged pistons, twin Weber 45 DCOE carburettors, a special cylinder head and camshaft, larger inlet and exhaust valves, reworked crank gear, a 5-litre oil sump with a modified oil pump, a competition clutch and competition exhaust. The main difference between the race and rally engines lay in their exhaust systems and camshaft profiles, the rally engine's camshaft being designed to improve mid-range torque.

In later years the Weber carburettors were replaced by Kugelfischer fuel injection, and power output rose to 220bhp. This was achieved at around 8,000rpm, which would have made a dry-sump lubrication system essential.

In February 1969, a road test of a Group 5 2002 with 180bhp at 7,000rpm, an 11:1 compression ratio and two Weber 45DCOE carburettors recorded a 0-62mph standing start time of exactly 6 seconds.

Race and rally parts

Some of the parts listed above were for road use, while others were restricted to race and rally applications. Alpina also produced many special parts for race and rally purposes only. The known items are as follows:

adjustable accelerator pedal
adjustable spring platform front struts
alloy clutch (215mm and 228mm)
aluminium knee supports, left and right
Bilstein racing front damper inserts

The tuned 2002

Bilstein racing rear dampers
BBS lightweight cross-spoke wheels, 3 piece, 5.5in to 11in
brake pads (Ferodo)
brake shoes with bonded linings
crossflow radiator with expansion tank

Thorough engineering and meticulous preparation were always the hall-marks of Alpina. The pictures opposite and above of their workshops date from the early 1970s.

driver's foot rest (clutch foot)
dry sump oil pump
electric fuel pump (carburettor and injection)
electronic cut-off switch (Digitron GRB 6)
front spoiler
full harness seat belts (Kangol and Britax)
high-compression forged pistons
high-pressure oil cooler
high-pressure oil pump
lightweight clutch release bearing
lightweight connecting rods (24mm and 28mm widths)
lightweight flywheel
polished and shaped inlet valves (46mm and 47mm)
polished and shaped exhaust valves (39mm)
progressive-rate road springs
racing camshaft (320° and 324°)
racing exhaust manifold
racing exhaust system
racing seat with head restraint
rally exhaust manifold
rally exhaust system
rally lamps (Cibie and Hella)
rally seat with head restraint
rollover cage for race and rally (ONS approved)
special rocker arms
stronger valve springs
shield for sump and gearbox (rally use)
titanium connecting rod nuts and bolts
twin-plate clutch
wheelarch extensions (9in front, 11in rear)
wheelarch extensions (to suit maximum 8in wide wheels)

The tuned 2002

AC Schnitzer

Like Alpina, AC Schnitzer had its headquarters in Bavaria. However, the Freilassing-based company had not always been involved with tuned BMWs: it had actally been established in 1934 as a truck repair workshop and driving school, progressing to take on various car dealerships over the next 30 years.

A Schnitzer head.

The Schnitzer association with high performance actually began in 1963, when Josef Schnitzer, son of the founder, started racing a Fiat Abarth 750. A year later, the association with BMW began when his brother Herbert opened a dealership for the marque in Freilassing, and Josef not unnaturally started racing BMWs. In 1966 Josef driving a BMW won the German saloon car championship, and a year later the two brothers founded the Schnitzer racing team.

Schnitzer's contribution to the history of the 2002 lay in the racing versions it developed. The company's 1969 Group 2 racing 2002ti was tuned to give 190bhp and ran in stripped-out form to give a maximum weight of 1,962 lb. That year, the even lighter (1,918 lb) Group 5 2002 boasted 205bhp; and there was also a 2002ti rally engine, tuned for torque rather than power, which gave 175bhp at 7,400rpm with a 10.7:1 compression ratio and two twin-choke Solex carburettors. Schnitzer took the European Championship with their 2002tii driven by Ernst Furtmayr.

After 1969, the focus of Schnitzer racing activity was on the BMW 2800CS coupé, with which Furtmayr took the European Championship twice more - in 1970 and 1971. However, work was continuing on the 2-litre four-cylinder engine, and in 1973 Schitzer unveiled their astonishing twin-cam conversion, which went under the name of the Type 20-4. This engine had a completely new cylinder head with four valves per cylinder, and developed 225bhp in rally tune for the 2002 saloon. Schnitzer also offered the engine for Formula 2 single-seaters, in which guise it offered 265bhp with carburettors or 270bhp with fuel injection.

The tuned 2002

Schnitzer found ways of getting even more power from the 2-litre engine. With Kugelfischer fuel injection and a 10.8:1 compression ratio, but still using the standard bore and stroke, they squeezed 292bhp at 9,400rpm out of the Rodenstock racing 2002 in 1976. This heavily-bespoilered car was allegedly good for 161mph. A further developed turbocharged engine gave as much as 600bhp in Group 5 cars for 1977, and in addition Schnitzer developed a 1.4-litre turbocharged 2002 saloon racer for Albrecht Krebs.

The tuning companies usually made sure that their handiwork did not go unnoticed! This is the Schnitzer twin-cam conversion of the 2-litre BMW engine.

GS-BMW-Tuning

Almost forgotten today, but an important name in 2002 racing during the 1970s was GS-BMW-Tuning. This company was an offshoot of the BMW dealership established in Freiburg by Gerhard Schneider, and opened for business in 1970. Within a year, Schneider could claim that 10% of all the new BMWs he sold were modified, and among them of course was a good proportion of 2002s.

For road cars, GS-BMW-Tuning offered a wide range of products from gear knobs upwards, but their most interesting development was a 16-valve cylinder head which became available in 1972 or 1973. Fitted to the 2-litre engine, this gave 285bhp. GS racing 2002s scored a number of victories in the early and middle 1970s, but the company was severely hit by the oil crisis at the end of the 1970s, and in 1982 Gerhard Schneider decided to call it a day and close his tuning business.

Chapter Seven

Buying a 2002 today

There is a wide choice of 2002 models available worldwide. However, even though the English-speaking countries (USA, UK and Australia) were strong markets for the 2002, not all models were available in all territories. In the USA, for example, only the 2002, 2002 Automatic and 2002tii were available. In the UK, the choice was slightly wider, embracing 2002, 2002 Automatic, 2002tii, Touring and Targa models, plus just 51 Turbos, the latter all with left-hand drive.

Even in Germany, 2002 models are now becoming a rare sight. Since the end of the 1970s, the Germans have been scrapping 02 range cars at the rate of 30,000 a year, and one result is that the majority of 02 series cars (including 2002s) nowadays are probably based in the USA, Japan and the UK. There are probably around 10,000 02s in the USA, 5,000 in Japan, and a further 5,000 in the UK.

Availability and prices

The single-carburettor version of the 2002 has excellent performance even by today's standards. There are plenty around to choose from, and the model's simple running gear and reasonable maintenance costs make it a suitable classic car to run as everyday transport. There is still a good supply of new parts from BMW and 02 specialists, as well as a plentiful supply of used parts. In short, there are few drawbacks to this model unless you desire a more potent 02 such as the ti, tii or Turbo.

Consider this: the 2002 has a top speed of 107mph, accelerates from 0 to 60mph in about 10.6 seconds, and returns between 25 and 30 miles to the gallon. An MGB GT has a top speed of 101mph, takes 13.6 seconds to accelerate from 0 to 60mph, and uses fuel at a rate of around 23 miles to the gallon. Even the Mark 1 Lotus Cortina tested by Autocar in 1970 took 11 seconds to reach 60mph from rest and showed a top speed of only 104mph.

Buying a 2002 today

The 2002ti with its twin side-draught Solex 40PHH carburettors was available only up until 1972, when it was replaced by the fuel-injected 2002tii. It was never available in the UK or the USA. The 2002tii model is the most popular of the 2002 range on account of its startlingly high performance, the excellent mechanical fuel injection providing a power increase of nearly a third as compared to the standard car. The cars are in good supply, and it is reasonably easy to find examples in the UK and the USA. Surprisingly, however, they are difficult to find in Germany because BMW exported most of those built. You often have to pay around twice the price of a standard 2002 for a nice example of a 2002tii or Targa-top cabriolet. Full cabriolets and 2002 Turbos can command even four times the price of a good 2002 or twice the price of a 2002tii. The hatchback Touring models seem to be of less interest to 2002 enthusiasts, although they are undeniably useful and practical vehicles.

It is wise to consider why you really want a 2002. If your requirement is for an everyday car to use often, you may do better buying a single-carburettor model. The torque characteristics and pulling power are greater at low engine revolutions, to the extent that after slowing for a sharp corner a 2002 will accelerate away in third gear whereas a 2002tii will need second. Once the 2002tii gets over 4,000rpm, however, the dust flies and you can really get a move on. With a top speed of 116mph and a 0-60mph time of 8.3 seconds, this model is no slouch; in fact a 4.2-litre E-type Jaguar was less than one second quicker over the 0-60 sprint. Surprisingly, most people find the fuel consumption of the injected models better, and an average of 35mpg is not unheard of.

Finding a 2002

Most classic car magazines carry a few advertisements for these cars in their back pages, but other good leads to suitable cars will be found through the BMW clubs and the specialist 02 series registers that exist in many countries. In the USA, the BMW Car Club of America is the one to contact; in continental Europe, BMW Club Europa can provide addresses of the various national clubs; and the UK has both the BMW Car Club (GB) and the BMW Drivers' Club.

What to look for when buying a 2002

The 2002 body shell is of monocoque construction, using metal pressings welded together to make a strong structure. The designer allowed a factor of safety for the strength of the shell, so do not panic if

Buying a 2002 today

Find this plate on the right-hand front inner wing, check the VIN against the lists in Appendix B, and then have a look in Chapter 2 or 3 to establish what the car would have been like when it was new. This is a good starting-point when buying; many cars have been modified, and not always sympathetically.

you see a little rust. However, rust does take its toll and, if neglected, the 2002 bodyshell rots away just as quickly as that of any other steel-bodied car of the era. The cabriolets and hatchback Touring models are more vulnerable to rust than the saloons.

It is always advisable to have a good look under the car when contemplating purchasing a 2002. Pay special attention to the mounting points for the rear suspension subframe, around the rear inner sill area. The rear subframe is bolted to the bodyshell with Metalastic mountings (rubber sandwiched between metal), and the rubber can perish and even break away from the metal. Also look where the rear of the differential bolts up to the shell: the body around this mounting can sometimes rust and the rubber mountings for the differential can perish with age. New rubber mountings can transform the handling of a car.

It is also advisable to check the metal box section above the rear springs, although this is difficult to see unless the rear wheels are removed. In the 1970s, rust-proofing was in its infancy, and usually there was no rust protection inside this box section, where the rear springs and dampers are mounted. Water gets in and runs between the original spot welds; you will often see a line of rust, visible from within the boot area, where the spring mounting panel joins the inner wheel housing. In extreme cases, the top of the spring can disappear into the underside of this box.

Whilst looking in the boot area, it is worth lifting up the spare wheel covering board, removing the spare wheel and looking at the condition of the boot floor panels. It is not unheard of for the spare wheel to drop out! If you suspect that the area around the fuel tank is rusted, you can either inspect this area from the underside of the car or remove the covering board, using a small screwdriver to remove the self-tapping screws which hold this panel in place.

In the same area, check the fuel tank for leaks. The fuel tanks are made from two sheet metal pressings which are electrically seam-welded together. The tank was bolted to the boot floor, using sponge foam as a gasket; unfortunately, this sponge was not "skinned" like the door and boot seals, and for that reason it absorbs water. This water penetrates the fuel tank seam and causes leaks when the tank is more than half full. If replacing the fuel tank on a 2002, it is advisable to throw away the original foam and replace it with suitable body sealer.

Sills can be a problem, and it is sometimes very difficult to know whether or not they have been replaced, either with cheap pattern oversills or with the much better and heavier original-specification panels. Strangely, both types are available from BMW! The oversill panels combine the front part of the rear wing and the outer sill panel under the door. The genuine sill panel, by contrast, consists of the outer sill panel under the door area and the strengthening member for the rear subframe mountings, the latter hidden behind the rear wing when the sill is fitted.

Pattern sills can conceal damage like this. Check carefully!

Some unscrupulous owners and restorers use an oversill to hide structural rust damage to the rear section of the strengthened original sill. From the outside, the sill looks perfect, but a year or two down the road the car will start to handle strangely and wander about the road. You can sometimes ascertain if the sills are original, or replaced with original, by checking if you can see the join between the rear wing and the sill. On original or properly repaired examples, there should be a join line just under the rear of the door. As the oversill panel is a combination of the two panels, there will not be a join.

Floor pans rust at both the front and the rear, but it is hard to check for problems because the floor is covered with heavy bitumastic pads which deaden vibration and road noise. The floor pans have four large-diameter plugs which were used for draining the bodyshells in the factory after dipping. These are a rust-trap and, if you are replacing a floor pan, it is best to leave them out. The replacement floor pans available today have the pressing for the plugs, but the metal is not cut out. If you are a stickler for originality, you can of course cut out the hole and fit drain plugs to original specification.

There is often rust trouble where the sills join the front bulkhead and the footwell sides. This results in water leaks, sodden carpets, and eventually yet more rust. There is also a plate on the outside front corner of the floor pan which was used in the factory to support the bodies when they were transported around the factory. These have no use now and are often better left off, as they are another rust trap.

The front wings bolt on, although they are lead-loaded at the front joint above the headlamps. They tend to rust from the inside out, just above the direction indicators and close to the door where the inner part of the wing is joined to the outer part. On poor examples of 2002s, you will find that when the front wings are removed, the inner wing strengtheners (sometimes known as front wing mounting bands) are rotten, along with the bottom section of the A-post. It is important when looking for a good example of a 2002 to put your hand up behind the front wings and check for crumbly metal.

Many body-shell problems are hidden. This pictures shows typical rust damage on the front inner wing of a **2002**. Few cars escape this kind of damage completely.

The front panels are very much prone to rust, too. The main reason is the rust trap at the front where the top of the front wing adjacent to the direction indicator lamp joins the front panel. This causes bad rust and perforated metal to the side by the headlamps. This part comes as part of the outer front panel as well as the complete front panel assemblies. The other area of the front wings prone to rust is the cross-member under the radiator where the overflow pipe is: sadly, the original overflow pipes were too short and disgorged their contents onto the metal of the car body instead of onto the road.

Where the fresh air for the heater comes in through the bonnet and then through the bulkhead to the heater, there is a box to isolate the engine compartment from the heater and interior of the car. Water from rain and car washing can come through the vent in the bonnet and lie in this box. Three rubber "socks" act as one-way valves to drain any water from this area while isolating the air for the heater from fumes in the engine compartment. Many owners do not realise what these rubber socks are for and let them become blocked with leaves and dirt. Water then has nowhere to go, and either lies in the bottom until the bottom rusts out, or overflows into the heater matrix and through to the interior. Either way, you get wet feet!

Buying a 2002 today

Some body problems will be only too apparent, as in the case of this **2002 TARGA**.

Wheelarches can suffer badly. In this case, the outer skin has been cut away to reveal that rust has taken a firm grip underneath.

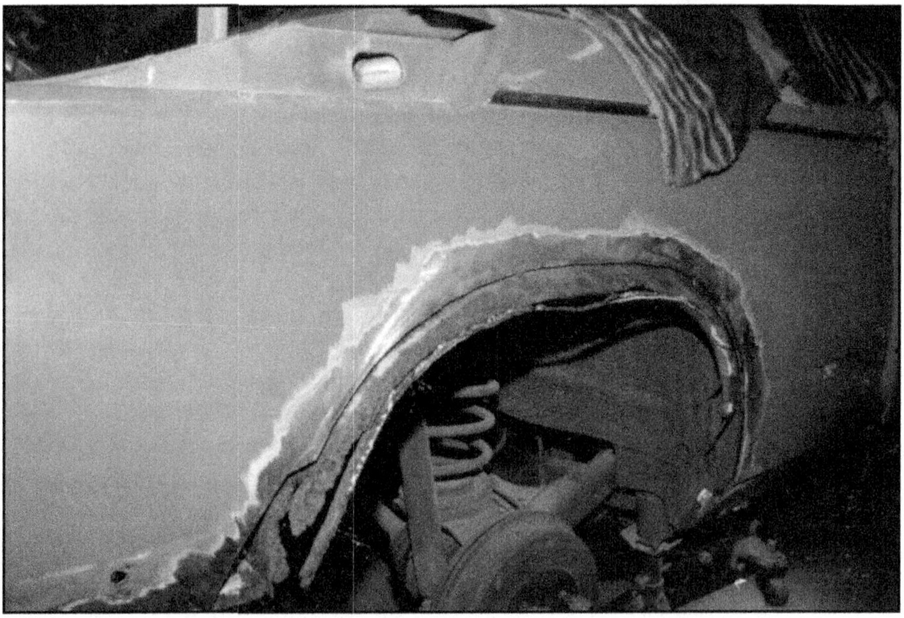

Doors are a rust trap, and the bottom frame of the door corrodes were the outer skin joins it. Doors are repairable, as repair parts and skins are available from specialists. However, it is worth considering new genuine BMW-made doors, as the cost may be less than repairing the old ones.

Cabriolets have even worse problems than the saloons. Be very wary of buying a "fully restored" full or Targa-top cabriolet if there are no photographs of the work done or bills for substantial repairs to the sills, rear wheel housings and floor areas. Most of the water which ran off the fabric hood was diverted by a rubber seal to run down the outside of the rear wing. Unfortunately, these seals do not stay very flexible for long, and some of the rain water runs down inside the wing to collect in the box that supports and stores the folded hood frame on either side.

Rusty rear wheel housings are a common **2002** problem.

That was just the start of the problem. Baur provided a small metal drainage pipe connected to a plastic hose (about the size of a fuel line) which extended down to the sill area. If the owner did not clear the 3mm hole at the bottom of the hood frame box regularly, the box would fill with rain water and overflow into the passenger compartment. Water would then lie on the floor under the rear seat. Even if the owner did keep this hole clear, there was still little chance of keeping the cabriolet watertight because the original plastic drain pipe ended inside the sill area, from where water was expected to escape through the very small slots on the underside of the sill panels. As these slots quickly became blocked with debris, water accumulated in the sills and promoted rust. Eventually, the inner sill strengthening panel and the inner sill itself rusted through, just ahead of the rear seat. Even nowadays, you can sometimes hear water sloshing about in the sills when you drive a cabriolet, especially under braking and acceleration.

The rare full cabriolets do suffer from a little scuttle shake. This can be seen when travelling over bumpy roads and looking at the boot lid in the interior mirror. The boot lid appears to shake or move. The Baur Targa-top shell is more sturdy, and there should be no problem with scuttle-shake as the permanent Targa bar helps to preserve torsional stiffness.

Buying a 2002 today

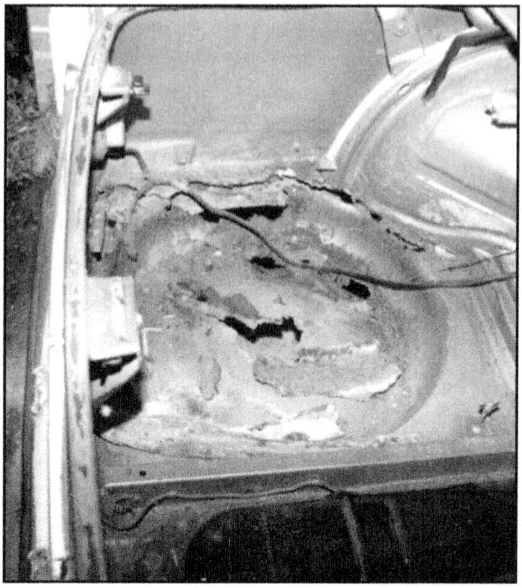

Top: Check carefully in the spare wheel well. Corrosion can get this bad.

Right: Beware of horrors lurking under the rear seat. These pictures show serious corrosion around the rear subframe mounting points on a **2002 TARGA** bodyshell.

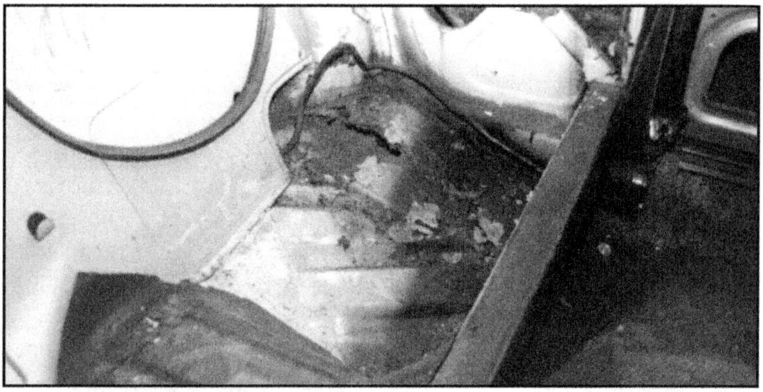

Buying a 2002 today

The Touring models are well worth inspecting closely for rust and distortion. Not having a rear bulkhead like the saloon models, they have an extra metal framing between the two rear wheel housings. The welds on each side often crack, which affects the torsional stiffness of the bodyshell. Major repairs to the Touring model bodies can be expensive because rear wings, rear panels and boot floors are no longer available. Repair panels are available for the rear wheel arches, but they are expensive when compared to the saloon arches because they are hand-made. Rear hatches are also expensive and suffer from the dreaded rust in the same way that the 2002 doors do.

The 2002's engine is a first-rate unit which did much to rebuild BMW's reputation in the 1960s. A well-maintained engine should reach at least 100,000 miles before it needs major attention. Problems are caused by neglect. When buying, look out for oil leaks, dirty oil and signs that the oil filter has not been replaced regularly. Check, too, that there is anti-freeze or rust inhibitor as well as water in the cooling system. Except for the fuelling, all 2002 models have basically the same engine. The camshaft, also shared with 1502, 1602 and 1802 models, is the same for all models of 2002 including the 2002 turbo; the single exception is the very late US-specification 2002.

Camshafts do not wear as much as those found in the big six-cylinder BMWs. Noisy engines are more likely to have worn rocker shafts, worn rocker bearings, or both. Valve stems, valve guides and valve oil seals all eventually wear, giving a tell-tale sign by producing clouds of unburnt oil smoke on a trailing throttle after running at high rpm. The reason for this is that the quantity of oil being pumped from the oil pump to the top of the engine at high rpm is greater than can drain back into the sump (until it reaches a certain level). As the 2002 engine is canted to the exhaust side, the valve guides and seals are surrounded by oil. When you lift off the throttle pedal, the supply of air and fuel that the engine was breathing is cut off and the engine tries to suck air from wherever it can. Air and oil are sucked down the valve stem and immediately pass into the exhaust port, from where they enter the exhaust manifold and finally leave the exhaust system as a cloud of blue smoke.

This small problem is not worrying from a mechanical viewpoint, as a little oil makes a lot of smoke if it is not burnt. But with the increase of environmental concerns in the 1990s, it does nothing to assist the classic car enthusiast's reputation! Listen out, too, for noisy timing chains, and watch for oil pressure lights that stay on at tickover.

The bottom end of the engine is strong and gives few problems if it is fed clean oil. To check for worn piston rings and general engine wear, remove the oil filler cap while the engine is running; if a great cloud of fumes is pushed out, there is a problem. If the tickover also becomes smoother with the oil filler cap removed, you can consider the engine will need a rebuild in the not too distant future.

Fuel systems are the main differences between engines fitted to the 2002 models, and there is no doubt that carburettors are less complex than fuel injection systems. The main problem found on high-mileage 2002s with the single-choke carburettor is that the carburettor butterfly spindle has worn, causing air to leak in and upset the tickover.

The Kugelfischer fuel injected 2002tii engine is more complex, and it is this that discourages people from buying a 2002tii. The system is excellent, and if understood and set up properly can be used for a long time with minimal attention. Unfortunately, neither the BMW workshop manual nor its Haynes equivalent explains the set-up procedure completely. Fortunately, clubs and specialists can often supply bulletins and details on how to maintain and set up the injection system.

When there are problems with the fuel injection of a 2002tii, the first component to be blamed is often not the cause of the trouble. If buying a 2002tii which does not have an even tickover or gives signs of black smoke on hard acceleration, look for worn linkages. This may be part of the problem and is fairly easily and cheaply rectified. It can also be a good bargaining point when negotiating the purchase! Because the mechanical injector pump is lubricated by engine oil, it can suffer wear if the oil is old, thin or dirty.

Another pointer to whether or not the system has been looked after properly is the presence or absence of an aluminium canister fuel filter by the left hand side of the radiator, in the fuel line just before the fuel delivery pipe to the mechanical pump. Many owners and garages fit a plastic in-line fuel filter instead, and as this type of filter was suitable only for carburettor models its presence indicates that the car has recently been serviced by someone who is not knowledgeable about the tii variant of the 2002.

Several different transmissions were fitted to the 2002s, but the vast majority of cars had four-speed Getrag manual gearboxes. Although five-speed close-ratio boxes were available as an option, few were supplied because of their high cost. The 2002 Automatic was fitted with a ZF gearbox.

On manual transmission cars, the gear linkage can become soggy, with excessive side-to-side movement. The usual cause is perishing of the mounting blocks which isolate the gear lever linkage from the gearbox. However, it is not a major problem and can be fixed cheaply and quickly.

Buying a 2002 today

Fortunately, engine bays do not often look this bad; this car had been standing in a field when it was rescued! Never underestimate the amount of work necessary to bring something like this back to usable condition.

To check for wear on the clutch, put the car into top gear, put the handbrake on, and try to drive away. If the car stalls, the clutch should be sound, but if the car stays where it is, the engine still runs, and you can smell burning, stop the test quickly: the clutch is on its last legs. Next, pump the clutch pedal six to ten times, and if it jams half way down its travel there is a blockage in the flexible hose to the slave cylinder which is making it act like a one-way valve. Have a look at this hose; if it is made of black rubber, replace it with the newer clear plastic type. If you find this problem on the clutch flexible pipe, it is advisable to check the brake flexible pipes as well for the same problem may exist with these.

Although gearbox internals have a good reputation, some cars fitted with the earlier coarse-spline output shaft eventually suffer from worn splines. This can be solved only by fitting a new output shaft, which can be expensive. The new shaft will be a fine-spline shaft, so you will have to buy the drive flange as well. The Borg Warner type synchromesh fitted to later gearboxes is fairly robust, but there may be some wear on five-speed and early four-speed gearboxes which used the Porsche-type synchromesh. Some owners have replaced the original four-speed gearbox with an overdrive five-speed type from the later four-cylinder E21 3-series; this was never an original option on the 2002, but it is a worthwhile modification.

The ZF transmission fitted to the 2002 Automatic is strong and reliable. As long as it changes up and down the gears easily, it is likely to be sound. However, if it hangs on to a gear too long before changing up, check the quantity of automatic transmission fluid. This must be done with the engine running and the gear lever in Park - a fact which many owners seem not to know.

Moving further down the driveline, the rubber coupling to the propshaft and the propshaft centre bearing both deteriorate with age and eventually require replacement. Look for cracks in the coupling, and listen for noise from the centre bearing. These problems are usually apparent on 2002s which have not been used for some time.

The main problem affecting differentials is bearing noise, but do check carefully because noise from the centre bearing can be transmitted along the propshaft, both to the rear axle and to the gearbox. It is not unknown for an owner to replace both the gearbox and the differential, only to find that the noise is actually coming from the propshaft centre bearing.

There was of course a limited-slip differential option for the 2002, which was often specified with the close-ratio gearbox. However, it is difficult to be certain whether a limited-slip differential is actually fitted.

There are two ways to check. One is to remove the rear cover from the differential and look inside. Limited-slip types contain an insert that looks like a large tin can, as well as the crown wheel and four smaller bevel gears. The second way to check is to jack one of the rear wheels off the ground with a trolley jack, ensuring that the trolley jack's wheels are facing in the same direction as the car's road wheels. Start the car and try (gently) to drive it forward. If it moves, you have a limited-slip differential. If not, either there is a limited-slip differential which is not working, or the differential is the standard type.

The last major elements in the drivetrain are the driveshafts, which run from the differential to the rear wheels. These are very reliable and rarely show problems. Rear hubs can suffer from wheel bearing problems, though, especially if the hub nuts have not been tightened to the correct torque.

On the steering, look out for play in the joints. If the adjusting nut for the ZF steering box has only a small amount of thread protruding from the lock nut, assume the steering box will require overhauling or replacement soon. The 2002 turbo models were fitted with a "quick" steering box with a ratio of 12.8:1 instead of the 15.5:1 ratio of other models. Stiff steering can be caused by one of four things. The first is a partially seized centre track rod, usually found on 2002s that have been laid up for some time. The second is rusty and seized suspension top mounting bearings. The third is a corroded steering idler from which the lubrication has been lost; and the fourth and last is wide tyres.

Suspension problems cause bad handling and often make braking and steering problems worse. Front struts can be fitted with new inserts, but the rear dampers come complete. Both front inserts and rear dampers are best replaced in pairs. The chromed centre rod of the strut insert and rear dampers tends to go rusty after lack of use, and this rust then cuts the rubber seals when the car is eventually used, causing the fluid to leak out.

The 2002ti, 2002tii and 2002 turbo models had box-section rear trailing arms, whereas the trailing arms on ordinary 2002s were simple channel-section pressings. When new, the box-section trailing arms were much stronger than the channel-section type, but with age they rust from the inside out, and are often found to be much weaker than the standard 2002 trailing arms.

The brakes, in general, are very good. Four-pot front calipers are used on all but the very early 2002 models, the 2002ti and 2002tii models having larger calipers and larger diameter discs. The Turbo models have the same type and size of caliper as the 2002ti and 2002tii models, but these are wider to allow for the ventilated brake discs. All the calipers fitted to the 2002 models can cause problems, especially if the brakes are not used often. The caliper pistons seize in their cast iron housings, and this in turn gives uneven braking. Replacing these pistons can be a problem, especially on the standard 2002 models, as the casting between the piston bores is thin, and cracks easily if excessive force is used to remove the pistons.

Two remote vacuum servos are used on the right-hand drive cars, and left-hand drive examples are fitted with a single servo which is combined with the brake master cylinder. All but the very early 2002s had a twin braking system where the master cylinder was effectively two master cylinders in one. One half of the system operates a pair of opposing pistons in each of the front four-pot calipers, together with the rear brake wheel cylinders. The other half of the master cylinder controls the other pair of pistons on the front left and right calipers. The idea behind the system was to maintain some braking even if all the brake fluid was lost from one of the two parts of the system.

Normal brake bleeding is very difficult with a split system, and the job is much easier with a pressure brake bleeding system. A good check for servo operation is to empty the system of vacuum by pumping the brake pedal, and then to start the engine. The brake pedal should move down. If it moves upwards to press against your foot, there may be a problem associated with one or other of the servos. On right-hand drive cars, it can be difficult to tell which of the servos is at fault. The easiest method is to have an assistant operating the brake pedal while you place the palm of your hand over the air intake on the underside of the servos. The servo that does not suck your hand to it is the one that needs attention. ATE, the manufacturer of the servos, recommends that servos should be replaced or rebuilt every 60,000 miles.

If the brakes pull to one side or the other, the usual cause is seized caliper pistons, and overhaul or replacement of the calipers will be needed. If the car pulls to the left under braking, it is most likely that the brake caliper at the right-hand front is at fault; and vice versa.

Like the clutch flexible hose, the rubber flexible brake pipes 'heal up' inside, leaving an extremely small hole for the fluid to pass through. This problem is difficult to discover unless you remove the brake hose and try to blow through it. If you cannot easily blow through it, the hose needs replacing anyway; so cut through it and you will be amazed at how small the centre of the pipe has become. As you exert a lot of pressure with your foot when you apply the brakes, the fluid has little problem in being forced down this small pipe, but when you lift your foot off the pedal, the fluid takes a long time to seep back past the blockage. This problem is difficult to find unless you remove the flexible hoses for inspection. But this problem can sometimes be noticed as a "shimmy" as you accelerate again after braking. Handbrake adjustment is virtually impossible on the 2002s if the rear brake drums are even slightly worn.

Inside the car, worn seat covers are unfortunately a perennial 2002 problem. The original cloth-covered seats give the worst wear rate; the late 1960s and early 1970s was the era of jeans, and the rivets on their back pockets did nothing for the longevity of the seat covers. It is not unknown for the seat base and backrest covers to show serious signs of wear on a 40,000-mile example. A lot of the original trim materials (including the carpet) are still available, so it is possible (but not cheap) to have the seats restored to as-new condition.

Upholstery is a weak point, though this car has suffered unusually badly.

Dashboards on Model 71 cars and right-hand drive Model 73s often develop a crack, and it has long been incorrectly assumed that this is caused by a defective bodyshell. In fact, it is caused by prolonged exposure to sunlight and a weak spot in the moulding. Sunlight can also be blamed for premature rotting of the cloth at the top of the rear seat back.

US-specification cars have a number of differences from European models, and buyers of these cars need to pay attention to certain additional features. The biggest problem area is the exhaust emissions control equipment, which needs regular servicing both to meet emissions regulations and to prevent other maintenance problems. Some parts are also hard to obtain: at the time of writing (1995), these included the air pump and its mounting bracket, the check valve, the diverter valve, the EGR filter and valve, and the EGR lines. Late-model cars have completely different emissions control equipment, and some items are frighteningly expensive to replace: probably the worst is the thermal reactor exhaust manifold for 1975 models, listed at $2626.78 in 1995.

Many 2002s sold in the USA were fitted with air conditioning, and it is advisable to check that the compressor is functioning correctly. This is usually a York item, but many owners change it for a better Bosch rotary type when replacement is necessary. It can also be difficult to find replacements for the condenser, expansion valve, drier and evaporator unit.

Chapter Eight

Running a 2002

Most 2002 models are usable classic cars suitable for everyday transport, either for weekends and holidays or for commuting to work. Perhaps only the 2002 turbo and the cabriolet models do not come into this category. The high value of the 2002 turbo as a collector's car tends to limit the risks which owners are prepared to take with the cars, and many turbos are now used just for club events, concours and occasional weekend jaunts. The full cabriolet models also seem to be only used nowadays as summer transport and are rarely seen on the road, which is perhaps not surprising when only 200 were produced.

The 2002 and 2002tii models are as fast and economical as many modern cars, and as such are well suited to the modern motoring environment. The press of the time described the 2002, ti and tii models as "long legged" when compared to other cars of the same period, although by more modern standards they are not. In the 1990s we have all now become used to very high geared modern cars with overdrive gearboxes, and so a 2002 now sounds 'busy' and seems to rev highly when motorway cruising at around 70 mph.

Some 2002 owners who use their cars regularly and drive high mileages convert their cars to use the five-speed overdrive gearbox from the four-cylinder E21 3-series of the late 1970s and early 1980s. This gearbox in top (fifth) gives fewer revs per miles per hour and so saves on fuel and engine noise. Its only disadvantage is that top-gear acceleration is reduced; but of course a change down into fourth (direct top) gives the same gearing and accelerative response as with the standard four-speed box.

Visibility is superb, even when compared to modern cars, because of the large expanse of glass. Parking is easy as you can see the front and rear corners of the 2002 from the driver's seat. The aerodynamic shapes of many modern cars make it difficult to judge the position of the front and rear bumpers, but this is not the case with a 2002. Being "square" shaped and about as aerodynamic as a brick nevertheless does not give the 2002 poor fuel economy. In fact, 25 to 30 mpg (Imperial gallons) is the norm for the 2002 and the 2002tii can return anything from 20 to 35 mpg depending on the condition and setting of the fuel injection system. In Europe, many owners of Ford XR3i models are surprised to learn that the single carburettor 2002 is as fast as the XR3i and similarly the 2002tii has the same performance and fuel consumption as the Vauxhall/Opel Cavalier SRi and several other modern "hot hatchbacks".

Insurance

One consideration when deciding on owning an 02 as everyday transport is the insurance. Many insurance companies now specialise in policies for classic cars, and many of these include agreed value, limited mileage insurance. The cost of insuring a 2002 on this type of policy can be a fraction of the price of insuring an equivalent modern car. Be sure you insure your 2002 with fully comprehensive insurance as you may otherwise have trouble proving the value of the car to the third party from whom you are claiming in the event of an accident.

Some **2002s** are still in everyday use by their originl owners. This is a 1974 model which still has a very low mileage and still belongs to its proud original owner, who is now aged over 70.

Spare parts and maintenance

When compared to other classic cars and even some other modern makes and models, the availability of spare parts for the 2002 is astounding. Nearly all the important parts of the engine and running gear are still available, partly because the 2002 shares many mechanical parts with the newer model BMWs. It is nearly possible to rebuild an 02 bodyshell using all new panels, and body parts like front wings are as cheap, if not cheaper than newer small Ford, Vauxhall or Nissan models. BMW has recently made a commitment to the classic BMW owner by starting a department called "Mobile Tradition" that will look after the re-manufacture of much-needed parts that are no longer available from BMW or BMW specialists.

The 2002 is one of the last of the BMW models that is owner-friendly when it comes to servicing and repairs. Workshop manuals are readily available in book form and most of the tools found in the home mechanic's tool box will suffice for normal everyday repairs. If you are new to BMW 2002 models, it is always worthwhile to ask an expert who has driven many 02s to test drive your car. You may not realise that the brakes, steering, suspension or other items are not up to the normal 02 standards. Being used to driving many 02s, the expert will immediately be able to reassure you that your purchase is sound and up to specification, or to suggest areas of the car's mechanics that require attention.

Regular attention needs to be paid to oil and water levels. Many owners do not realise that the oil in an engine is a major factor in the cooling of the engine and that its level should therefore be regularly inspected. The coolant level is also most important as it is very easy to completely ruin a cylinder head by allowing the engine to overheat. If the temperature gauge goes into the red sector, move the heater lever to hot and check that hot air is coming from the heater. If not - stop the engine immediately. Check for steam appearing from the radiator overflow pipe. Do not remove the radiator cap before the engine has cooled down as you may be showered with boiling water. Investigate the reason for the loss of coolant and repair the cause of the problem before continuing your journey.

The **TOURING** models still make very practical everyday cars in the age of the hatchback.

Occasionally a small, regular loss of water from the cooling system can be put down to a crack in the cylinder head casting and an exhaust port. In normal driving the coolant loss is minimal, but when the car is driven very hard the crack enlarges and the coolant loss is excessive and causes overheating. As the water pump is belt-driven a careful eye needs to be kept on the tension of the fan belt, too, because looseness can result in a loss of efficiency of the water pump. This will often be accompanied by a screaming sound from the fan belt when the headlamps are switched on, requiring more current from the battery and alternator. The rubber bushes on the alternator mounting go soft and cause the alternator to tilt towards the water pump and in turn cause the belt to go slack.

Some items on the 02 models are best left to the experts when it comes to repairs and overhauls. The gearbox is fairly complex and many of its bearings have to be withdrawn with special pullers when the unit is disassembled. Unfortunately for the home mechanic, BMW used several different suppliers for the bearings, and some bearings which have the same dimensions nevertheless contain different numbers of balls. Therefore, to overhaul the gearbox you require a range of expensive bearing pullers to carry out the gearbox strip down. Similarly, the differential and steering box should be left to the experts to overhaul and rebuild.

Brakes are a puzzle to many 2002 owners as cars have a split hydraulic system. With this system, a failure in one half of the system (e.g. broken brake pipe) still leaves the car with adequate brakes. The problem is that the split circuit makes the system difficult to bleed. To bleed the system successfully, you need to use a pressurised brake bleeder to push the fluid through the system. Many types are available, from the cheap "Easi-Bleed" by Gunson to the expensive ATE pressure bleeder used by workshops. Pressure bleeding is much easier than trying to drive the brake fluid through the system by pumping the brake pedal. It is also less likely to damage the master cylinder piston seals.

Few owners go this far in making their **2002s** into vehicles suitable for everyday use ...

Laying-up a 2002

If leaving your 2002 unused for some considerable period in a damp environment (two to three months or more), you may find a whole collection of problems when you return. The clutch driven plate may have stuck to the flywheel, the handbrake may have jammed on, the steering may have become very stiff, and the propshaft centre bearing may have become noisy. However, such problems are not inevitable. Prevention is always better than cure, and there is a proper way to lay up a car.

You can prevent clutch problems by blocking the clutch pedal down with a suitable length of wood between the driver's seat and the pedal (use another piece of wood across the seat to spread the load). Leaving the car in gear with the handbrake off will ensure that it cannot roll and will also prevent the handbrake from seizing on. Some grease around the propshaft centre bearing and the two joints between the centre track rod and the steering drop arm and idler arm will stop the ingress of moisture which causes stiffness and noise.

Running a 2002

Altogether more practical is this conversion of a standard saloon, which aims to recreate the rare **2002 CABRIOLET**. The main visible difference is in the windscreen, which remains the standard saloon type instead of the lower and more raked convertible item. The conversion is known as the **Cabrio2**, and is available in Great Britain through Jaymic.

Running a 2002

Top left:
Many **2002s** still in everyday use have been quite extensively modified. This UK- registered example features the popular 'boxed wheelarch' conversion, which uses GRP panels.

Bottom left:
An unmodified **2002** turbo is perhaps not the best car to use every day. Most examples are now kept for club events and other special occasions. These two were pictured at the BMW Car Club's (G.B.) gathering at Brooklands in 1993.

Top right:
This apparently near-standard US - model **2002** has in fact been extensively modified.

Bottom right:
...and the owner of this one has perhaps gone a bit too far!

Running a 2002

Running a 2002

Useful tips

BMW 2002 owners tend to come up with the same problems over and over again, and we have been able to compile a list of tips to combat these problems. Here it is, in alphabetical order.

Alternator not charging:
Check that the brown earth wire is connected from the alternator to the engine block.

Blue smoke on the overrun:
This is a sign of worn valves, guides or valve stem oil seals (and it can be a sign of all three together). The engine actually burns little oil, but the blue haze from the exhaust can be annoying to drivers travelling behind.

Brake pulling:
Is normally caused by seized caliper pistons. Another possibility is that a rubber flexible brake hose has "healed up" inside and is acting like a one-way valve. It then allows the fluid to pass through under pressure, but not to return. This can also cause brake fade in extreme cases.

Brake servos not working:
With the engine off, pump the brake pedal 6-10 times to get rid of any vacuum left in the manifold. Leave your foot on the brake pedal and start the engine. If the pedal goes down slightly, the servos should be functioning properly, but if the pedal presses back against your foot, the servo or servos may be faulty.

Brake warning light stays on:
In all probability, the switch behind the handbrake lever will be at fault.

Car not level:
Check that the rear springs have not sunk into the box section above them. This box section was rarely rustproofed when new and rusts from the inside

out. The height at the front and back can be adjusted by using spacers of different thicknesses.

Clutch slow to respond:
Try pumping the clutch pedal half a dozen times. If it jams half way down its travel, the rubber flexible pipe to the clutch slave cylinder requires replacing.

Cooling:
Never run the cooling system without anti-freeze or a corrosion inhibiter. The copper radiator, alloy head and cast iron block make for a very corrosive mixture! Extend the overflow pipe to miss the front panel cross member or you may end up replacing the front panel sooner than you thought.

Brake warning light stays on:
In all probability, the switch behind the handbrake lever will be at fault.

Distributor. Which way does it turn?
Clockwise.

Dull headlamp reflectors:
The reflectors are silvered rather than chromed, so never wipe them clean with anything that may damage the silver surface. Make sure the rear plastic covering bowl is in place; it is there to keep out moisture.

Flickering instruments:
The flickering is caused by a bad earth. Check the soundness of the earth points (brown wires). They can be found by the battery, on the engine bulkhead (on the tii and turbo), behind the glove box, in the boot area under the petrol tank cover board and, last but not least, there is a large earth lead to the engine block.

Fuel injection problems:
Check fuel filters and fuel delivery back to the tank. Delivery, with the engine not running but the ignition

on, should be 850cc in 30 seconds. When setting the linkages always hold the linkage back, as if your foot was lightly on the accelerator pedal, to eliminate slack in any of the linkage ball joints. Always use the correct injection set-up tools. Do not let people who are not experts in the Kugelfischer system anywhere near it - you probably know more about the system than they do!

Fuel insufficient at the carburettor:
The mechanical fuel pump may not be the cause of the problem. Check that air is not leaking in through perished rubber fuel piping which connects from the fuel tank to the white plastic pipe that carries the fuel through the car.

Handbrake will not adjust:
You have replaced brake shoes, handbrake cables and the seized brake back plates and still the handbrake will not work? Replace the brake drums - a very small amount of wear on the drums will affect the adjustment at the handbrake lever considerably.

Headlamps failed, or only giving main beam:
This is usually caused by a bad contact on a fuse. Clean all of the fuse contacts, not just the headlamp ones. If that does not do the trick, check that a wire has not dropped off the back of the main light switch on the dashboard.

Rear hub nut. What is the size and how tightly should it be tightened?
The spanner size is 36mm and you must tighten it as much as you possibly can, preferably with a metre length of pole over your socket bar! The torque setting is beyond the scale of most DIY torque wrenches.

Starter motor will not turn:
If the cause is not a flat battery, the centre small trigger wire may have dropped off the starter motor solenoid.

Stiff steering:
If the car has not been driven much recently, the centre track rod ends may have rusted or the strut top bearings may have partially seized.

US-specification models:
There are some problems that are unique to the US-specification 2002 models. A very common one concerns the emissions air pump that needs to be kept in good working order; otherwise expensive replacement or repair will be necessary. The mounting bracket for the air pump can break if the belt tension is too tight. The late model EGR (Exhaust Gas Recirculation) system is more complex and can need expensive replacement parts if defective.

Weak running on number four cylinder:
This is usually caused by a leaking perished servo vacuum hose.

Chapter Nine

Beneficial Modifications

The standard unmodified models are perfectly adequate for most owners. Others, however, would prefer modifications to comfort, lighting, brakes, suspension, engine, and so on. Around the world you will find 2002s supercharged, 2002s fitted with M3 engines and even a 2002 that has been fitted with a modified 24-valve M6 engine. Some people just aren't satisfied with the standard production models!

As the 2002 models were popular with race and rally enthusiasts, there were a myriad of "go faster" and sports parts available if you could afford them. Five-speed close-ratio straight-cut gear sets are still available, with a choice of ratios, but at a price. Just the gear set alone costs around the price of a really nice 2002tii!

Obviously the modifications that you can perform on a 2002 are endless, so in this chapter we will examine some of the more popular types and try to give some idea of their benefits or (where appropriate) their drawbacks.

As cars become older, ideas that motor manufacturers incorporate into their new models filter through to the owners of the older models. Some owners then incorporate these changes into their own cars. One significant occurrence of this on the 2002 models is the wheels.

Wheels and tyres

When BMW introduced the 3-series models they used a wider version of the late model 2002 steel road wheel on the car. BMW also offered alloy wheels as an option and the aftermarket suppliers cashed in by offering relatively cheap alternative alloy wheels. This left a large quantity of second-hand original 5.5x13 3-series steel wheels on the market at very low prices. These wheels were purchased by the owners of all variants of the 02 range and by used car sales establishments to modernise the look of the earlier 02s and to fit wider wheels and tyres to the later

Beneficial Modifications

Top:
Beware of these late-pattern "standard" wheels. They do look good, but many cars have been fitted with similar wheels from the 3-series, which actually have a different offset.

Bottom:
Minilite-lookalikes on this UK-market car enhance the appearance.

examples. Unfortunately the offset on the 3-series wheel was not the same as on the 02 models, and with a full load the tyres would often rub on the front and rear wheel arch lip. These incorrect wheels can still be found today on 02s, as can the 3-series Mahle "cross wire" and "Alpina style" alloy wheels which can cause the same fouling problems.

The original 165SRx13 and 165HRx13 tyres are becoming more and more difficult to find, especially in America, where it now seems popular to use the 14-inch alloy wheels from the E30 3-series. With 13-inch and 14-inch wheels, you can increase the wheel width to 6 inches and use up to 205/60 profile tyres, but the wheel inset must be between 25 and 28mm if the tyres are not to foul the standard wheel arches. With 15-inch wheels you can use 205/50 tyres with 6-inch rims on the front and 7-inch on the rear. But remember, the wider the wheels, the harder parking manoeuvres become. Effects of "tramlining" also increase with wide wheels.

Suspension

Any modifications to the suspension system of your 2002 will be a trade-off between comfort and handling performance. BMW originally designed the 02 models as sporting saloons, so that the handling was good to start with. If you wish to improve the handling bias of the suspension, you have to appreciate that your modifications will make the ride and less forgiving.

Going to the ultimate extreme, a full race 2002 will have solid subframe mounting bushes, no rubber in the front suspension top mountings, and all the suspension joints will be rose-jointed or the rubber pivot bushes will have been changed to urethane-type materials. The suspension dampers front and rear will be extremely hard as will the springs. The body will be as close as possible to the ground to lower the centre of gravity. The tyres will be low profile and much wider than on a standard 2002. If you drove such a race car on normal bumpy roads and in town traffic, you would feel that you had completed the Mille Miglia after only an hour's driving! The steering would be very heavy and you would feel every bump vibrate through your body; most likely, the car would often "ground", as well.

So what you want for a road car is a compromise. Our suggestions are that you go no harder than 20% uprated springs. On height, go no lower than 1.5 inches less than standard and use good quality "sport" front strut inserts and rear shock absorbers. New rubber suspension

Beneficial Modifications

The right wheels can make a vast difference to the appearance of a **2002**. The alloy wheels fitted to this **2002 TARGA** were a factory option.

US enthusiasts today often fit the 14-inch alloy wheels from the **E30** models. This car also sports an aftermarket front spoiler made by Zender.

bushes will transform a 2002, especially if the original ones are still fitted to the car. If you still require less roll or want to get rid of too much oversteer you can fit a larger 19mm front anti-roll bar from the 2002 turbo; if you wish to improve this even further you can fit urethane bushes to the front and rear bars, or even fit the Alpina type rose-jointed anti-roll bars. Adjustable rear subframes are also available, and these allow you to adjust out any excessive rear negative camber caused by lowering of the car. This then improves grip and reduces tyre wear.

Engine performance

This is another area where there is a trade-off when you make modifications. It's all well and good having 200bhp, but is it usable? If you are only getting the power at 7,500 rpm, it's not much good for driving in traffic!

Sensible power increases can be gained by adding twin Weber 40DCOE side-draught carburettors, a maximum of 300 degree camshaft and a good tubular extractor exhaust manifold. However, do not expect as much smoothness and quietness with Weber DCOE (or the alternative Dellorto DLHA) carburettors as you get with the factory-standard 2002ti. The original twin Solex 40 PHH carburettors were very sophisticated pieces of equipment, and when driving a good original 2002ti it is difficult to realise that the car is fitted with twin side-draught carburettors. In the UK twin Solex 40 PHH carburetters do turn up for sale second-hand; unfortunately most of these carburetters come from the 2000ti models, which used a different linkage system; you will therefore need to make many modifications to get them to fit the 2002 models.

With right-hand drive cars, the four-branch tubular exhaust manifold is a problem. It is virtually impossible to design and build a good 4-into-2-into-1 system as the steering box and column get in the way. With left-hand drive cars this is no problem.

It is difficult to improve the power output of the 2002tii engine because tuning potential is limited by the amount of air you can get into the engine without resorting to the 4 butterfly type injection system as used on the Alpina A4/S. This set-up is no longer available from Alpina and second-hand systems are as rare as hen's teeth. Other systems are available from specialists, but we would advise against fitting 'slide throttle systems for road use.

Many camshafts are available for the 2002 models and work fine on the carburettor models. The 300 degree camshaft is the only

Beneficial Modifications

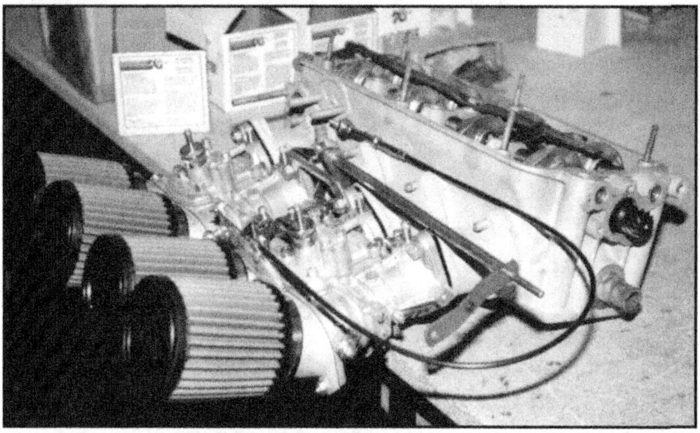

Many owners like to tune their **2002s**. This picture shows twin Weber 40DCOE carburettors in place.

Another twin-carburettor modification, this time also including a modified cylinder head and special air filters.

This modification may not be common, but it is certainly possible: this owner has fitted his **2002** with the engine from a later **M3** model.

camshaft (other than standard) that we have found works compatibly in conjunction with the injection system on a road car. Even then the injection system needs very careful setting up by a tii expert.

If you increase the performance of your 2002 engine up to tii power outputs of 130bhp, don't forget to convert the braking system to tii specification or better. If you go beyond 130bhp, perhaps to 150bhp or even higher, you should remember that even the tii braking system will not be adequate. For power outputs of this order, you should be thinking about fitting a 2002 turbo braking system or a similar brake upgrade offered by many 02 specialists.

Brake conversions

We cannot stress too highly the importance of good brakes. If you improve the performance of your 02, also check that the braking system is in first-class working order. Uprate the brakes to ti and tii specification if you increase the power to 115-150 bhp; 2002 turbo specification would be even better, and is essential with power outputs of more than 150 bhp.

Five-speed close ratio gearbox

Surprisingly, the 5-speed close-ratio gearbox that was offered as an option had a higher first gear than the standard four-speed type. In layman's terms, this means that the gears in the 5-speed gearbox are even closer together than if you included an extra gear between first and fourth in the four-speed gearbox. The five-speed close-ratio gearbox was an expensive extra, costing the owner rather more than 10% on top of the purchase price of the car; as a result, not many 02s had it fitted. Fitting the five-speed close-ratio gearbox to the 2002tii and 2002 turbo models is of more benefit than fitting it to the standard 2002 models, because the gearbox is better suited to the narrower power band of the more powerful cars.

Five-speed overdrive gearbox

Although the 2002 models were never fitted with a five-speed overdrive gearbox, the conversion is popular amongst owners who use their cars for motorway and long distance driving. The higher gearing of the overdrive fifth gear makes for quieter travelling and better fuel economy.

The only gearbox that is suitable to fit into the 02 body without major surgery is the five-speed overdrive gearbox from the four-cylinder E21 3-series cars (that is, the pre-1982 316 and 318 and the earlier four-cylinder 320 and 320i models). With the gearbox you will need the front

Beneficial Modifications

Boxed wheelarches made of GRP - seen here on a US enthusiast's much modified **2002** - undoubtedly enhance the car's appearance and can also allow for wider wheels and tyres

This aftermarket rear spoiler - seen on an early US-market **2002ti** updates the appearance but probably has no practical value.

For high-performance conversions, it is a good idea to fit the factory option ZF limited-slip differential.

part of the 3-series propshaft along with the clutch release bearing, clutch release arm and slave cylinder. It is also useful to pick up the gearchange mechanism at the same time. Because the 2002 front propshaft coupling has four bolt holes and the 3-series item has just three holes, the front portion of the 3 series propshaft needs to be grafted on to the 2002 propshaft. This is not a job for the amateur; the two halves must be properly mated and the propshaft balanced.

This is not the place to go into the intricacies of fitting the gearbox. There have been many articles in magazines about this conversion and any 02 clubs or specialists can point you in the right direction for further information.

Lead free conversion

When all about you are driving "green" machines, and the adverts on the television are trying to persuade you to trade your car in for a newer cleaner car, don't despair! You can convert your 2002 to run on lead free fuel - even on methane gas, if you insist. All you need to do is to have the exhaust and inlet valve seats replaced with seats made from a much harder material. The BMW valves are very good and do not need to be replaced. However, the price difference in the UK between 4-star petrol and lead-free fuel is too small to make the conversion cost-effective unless your cylinder head requires an overhaul anyway.

Cooling

Cooling is a big problem on the 2002 models, especially if all the parts in the cooling and heater system are not up to scratch. Electric fans can help, and it is possible to fit a larger radiator or even fit the radiator from the 2002 turbo. One of the simplest solutions to get the temperature gauge needle to stay out of the red is to play around with the different temperature thermostats that are available. The normal thermostat for the 2002 models is an 80 degree thermostat, but you can fit a 75 degree (as fitted to the 2002 turbo) or even a 71 degree type to make the engine run cooler. If your problem is an unusual one in that your engine will not run hot enough, you can fit the 83 degree thermostat.

Another valuable tip about cooling systems can help ensure a long life for the front panel on any 2002. If you replace the radiator overflow pipe with a pipe that is three inches longer, any coolant forced from the radiator will be directed on to the road instead of onto the panel itself, where it will promote rust.

Seats

Turbo and sports seats for the 02s have always been very sought after and there are very few for sale on the second-hand market. The answer is to find second-hand Recaro seats from the E21 3-series and recover them in the 2002 seat materials. The finished article looks very good and is extremely comfortable.

Four-headlamp conversion, spot lamps and rear fog lamp

BMW originally offered a four-headlamp conversion for the 2002 models. This used the headlamp assemblies from the 2500-3.0Si range of models and entailed cutting away the original 02 headlamp mountings. The four-headlamp grilles have now been remade, and better mountings for the four lamps are now available from most 02 specialists, as are copies of the original optional spot and fog lamp bars.

It is not common knowledge that a rear fog lamp was offered as an option and it is very rarely that you see this option fitted to 02s. Similar ones are still available and fit just under the rear bumper.

Map reading light

The interior light is very poor when you need illumination to help with navigation at night. The best solution to this problem is to buy one of the aftermarket map-reading lights that plug into the cigarette lighter. These are a good addition for any 02 owner who needs a map reading facility.

Long range fuel tank

Owners who regularly travel long distances in their 2002s often wonder whether it is possible to increase the fuel capacity from the standard 55 litres (12.1 Imperial gallons, 14.5 US gallons). A relatively simple solution is to fit the larger tank from a 2002 turbo, which will go in without major modifications. This will give you a capacity of 70 litres (15.4 Imperial gallons, 18.5 US gallons).

Electronic ignition

There are many systems on the market, some good and some not so good. Advice sought from other 02 owners is often the best policy if you want an unbiased opinion. Specialists will be selling their "favourite" system, but it may be their favourite purely because it has a larger profit margin than other makes! With 2002tii models, electronic ignition can be a particularly worthwhile addition because it will often improve the starting characteristics.

2002 turbo improvements

Back in 1974, turbochargers were in their infancy. Design changes over the years have produced turbochargers with more flexible reponses that give a more progressive power increase. Any 2002 turbo owner who uses his car for everyday motoring should at least consider changing to one of the more modern Garrett Air Research turbo chargers, especially if it is necessary to replace the original KKK type on the car as a result of a cracked housing.

Spare keys

This seems a really silly item to include under a chapter heading of 'Beneficial Modifications'. However, it is nowadays important to have a spare set of keys as the majority of key cutting emporiums do not keep key blanks in stock for the 02 models. Make a note of your key number in case you lose your keys. Sometimes you can find the key number for the door and ignition locks on a small 10mm x 3mm white label stuck to the back of the lock.

Headrests

In the 1960s, headrests (more properly, head restraints) were a novelty, but BMW did offer them as an option on the 2002 models. Three decades later, the chances of having a rear end shunt are greater, and as a result many 02 owners want to fit headrests to their early cars.

However, not all 2002 headrests are the same, as BMW changed the design a few times! So before purchasing a pair of second-hand headrests for your 2002, check the shaft diameter and the distance between centres of the holes or blanking plugs in your seats.

Beneficial Modifications

The original **2002** turbo was never a very easy car to drive. Replacing the original turbocharger with a more modern type can make it a much more tractable machine without sacrificing any of its formidable performance. This replacement turbocharger is made by Garrett.

Chapter Ten

Restoring a 2002

The simple title of this Chapter somehow makes the whole business of restoring a 2002 sound much simpler than it really is. Make no mistake, though: there are all kinds of problems and pitfalls waiting for anyone who actually attempts to do the job. If that were not the case, there would be no need for this Chapter at all!

Planning

Before you start upon the treacherous road called Restoration, you really need to find answers to a whole series of questions beginning with "if", "what" and "when". The planning stage is probably the most important part of any restoration. Get things wrong at this stage and you can ruin everything, possibly including your marriage! So the first question you must ask yourself is, "Why do I want to restore a car anyway? The answer to that may well be one or more of the following:

1. I want a hobby; I have a lot of spare time.
2. I want a car that is enjoyable to own and drive.
3. I want a car to show at classic car events.
4. I want a car so that I can join in with owners' club events.
5. I want the satisfaction of restoring my own car.
6. I want a car I can use that is different from all the Euroboxes on the roads today.
7. I want a car for posing, something that is different and will draw attention.
8. I want to make money by restoring a wreck and selling it for a profit.

All of these are genuine enough reasons for wanting to restore or own a classic car. But reason number 8 can be a problem. As your aim is to make money out of the restoration, you will be cutting corners; and because of the time and money you want to recoup when you sell the car on, you will have cut down the number of people who can actually afford to buy it. In fact, it is highly unlikely that you will recover the cost of a proper full-blown restoration. Far better, if you buy a cheap car that needs restoration, to sell it on straight away and settle for a small profit. Then use the money to buy the car you really want.

If you are still convinced that you want a classic BMW 2002, and wish to restore it, read on.

Benefits of a classic BMW

In addition to the eight reasons outlined above, there can be other benefits of owning a 2002, especially as an alternative to a modern new or nearly new car. These can be:

1. Cheap classic car agreed-value insurance policies.
2. You can pretty well ignore depreciation, and there is a good chance that, in 10 years' time, your classic BMW will be worth more than it cost to restore.
3. If you restore or rebuild a car yourself, you will know it inside out. So if it breaks down, you will know what to do!
4. The classic car will be fun to drive.
5. If the car is a popular model, like the 2002, there will be a plentiful supply of second-hand spare parts.
6. You will even be helping the environment, by recycling an existing product instead of using energy and resources to make a new one. Remember that old BMWs are just as recyclable as new ones!

Which model?

If you have decided that you definitely do want a BMW 2002, the next question is which model to choose. You need to think carefully and logically about the questions below, and to be careful not to fall into the trap of buying a car that you can not afford to run or maintain. Don't let your heart rule your head!

1. What budget do I have?
2. Do I want a standard car or a performance version?
3. Will I use it as a first or second car?
4. Can I garage it?
5. Has it got to be a family car?

6 Can I really afford the insurance?
7 How long can I spend restoring it?
8 Do the rest of the family agree with me?
9 How much of the work will I do myself?
10 How much work will have to be carried out by professionals?
11 Do I have the room to restore the car myself?
12 What will I use the car for?

You will probably think of a few more questions of your own. Above all, though, the Golden Rule is DO THINK FIRST.

By now, you should have some idea of the car you would like and the car that would be best suited to your needs. You might think that the best solution is to purchase a restored car. All well and good, but you need to remember that properly restored BMWs do not often come up for sale. Most people who have restored cars or had them restored to a high standard do not want to sell them; they want to keep them!

Restoration work will usually uncover all kinds of horrors lurking underneath the paint, but there is no substitute for the peace of mind which a thorough stripdown can bring.

Bodywork restoration

If you intend to use the car on the road after you have bought it, carry out a thorough mechanical inspection and deal with any repairs here before you even think about doing any restoration on the bodywork. Once past that hurdle, you can begin to think seriously about the next stage.

However good or bad the body may appear, you should make a very careful inspection as a first step. Steam clean the underside, wash the outside thoroughly and remove carpets, interior trim panels, and anything else which is going to hinder that inspection.

Don't start by ripping everything apart and dismantling the car! You need a more reasoned approach, and one way of doing it is to make notes as you examine things. A useful aid is a clipboard with a sheet of paper divided into columns for:

Remove (and description) - Refit - Repair - Replace - Paint - Notes

You can note down the part or area of the car in the first column, and simply put a tick in the relevant column after that.

Restoring a 2002

Start your inspection at one end of the car and begin by looking critically at the body panels alone. For example, on a 2002 the front panel is a vulnerable area. Ask yourself if it needs replacing. Will you need a complete panel or just an outer panel? What will you have to remove to replace the panel? And will you need any special tools, such as spot weld cutters?

Fill in your sheet for each panel, putting a description and a tick in the appropriate column. Do not forget to list the strengthening panels and the panels underneath the car. Now go round the car again, looking at chrome mouldings, handles, lights and so on. Note again the description and whether the part needs repair, replacement or painting. Be very critical on chrome parts, as what may look like a very small blemish or scratch now when the car needs paint or bodywork restoration could look terrible on a bodysell that is freshly painted and immaculate.

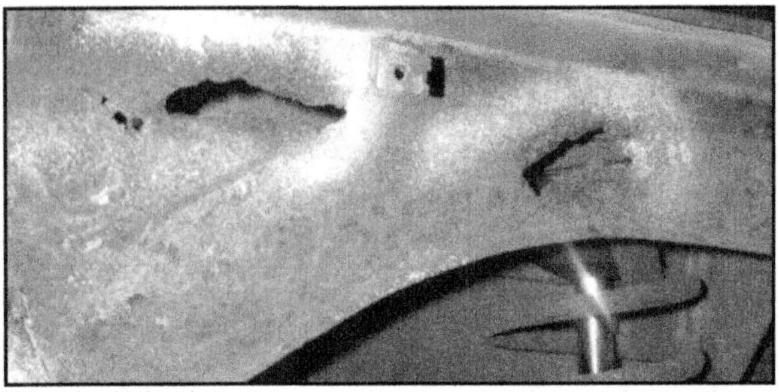

Areas like this front inner wing must be dealt with properly. The rust must be cut out and new metal welded in.

It is advisable to go round the car for a third time, this time viewing the glass and rubber seals. Again, a small scratch will look a lot worse when everything around it has been restored to pristine condition. Now you have a detailed list of panels and parts that you need to replace or repair, so the next stage is to find out the prices of those parts. You also need to check on their availability, and whether repair sections could be used instead of complete panels.

Armed with your first list and your parts prices, you again need to make some decisions and ask yourself some questions.

1 Do I have the tools and equipment necessary to carry out the body restoration myself?
2 Do I have the skill to carry out the repairs?
3 How long is it going to take me to do the restoration?
4. Should I get the bodyshell restored professionally?
5. Can I save labour costs by carrying out some of the work myself?
6. Would I be better off to get a lesser model with a good bodyshell and transfer the mechanical elements from this car to it? (For example, you could put the mechanical parts from a well-used and rusty 2002tii into a sound 1502.)

More costing comes next. Let us assume that the rear wing (or quarter panel) of your 2002 is rusted through on top of the wheel arch and at the back below the bumper corner. You have various options. You can repair the wing using a wheelarch repair panel and an under-bumper repair panel, or you can replace the complete rear wing. If you are doing the repairs yourself and can both weld and repair bodywork, your best option is likely to be the repair panels. But if you are having to pay somebody to carry out the work, you may be better off having the complete wing replaced, because it takes less time to replace and refinish a wing than to insert the two replacement panels invisibly and refinish the wing.

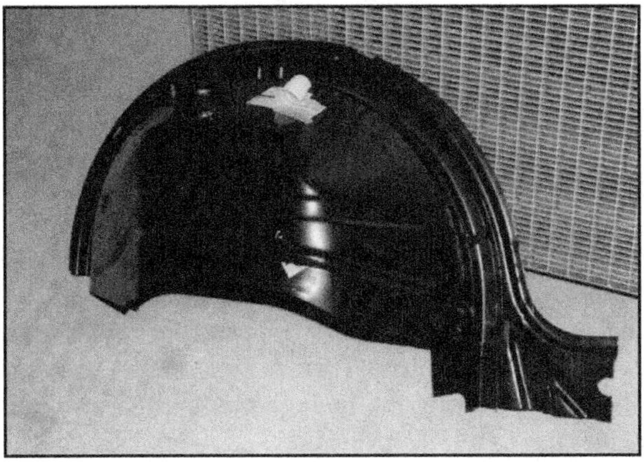

In many cases, fitting a complete new panel is more cost-effective than patching a rust-damaged old one. This is an original-specification BMW rear wheelarch panel.

Make a calculation like this for everything on your list. Add it all up and you will have some idea of how much work is required, how much it will all cost, and maybe even how long it will all take. It is always useful to set yourself a target date, such as a particular club meeting or event, as this can spur you on when things get difficult - as they will! To give you some idea of times, an average professional restoration on a 2002 takes aound 500 hours to complete; and that is when the work is being done by someone who knows a great deal about the cars and their problems. Allow more time than you think you will need to complete the job.

Carrying out a successful restoration can be rewarding, and can be one of the most satisfying things you will ever do. But be warned: it can turn out to be a disaster. Boxes of bits and half-completed restoration projcts do not fetch half as much as complete cars for restoration!

Do I go concours?

Some 2002 enthusiasts want to own a concours-winning car. If you are among them, you need to come to terms with a few facts. The first is that you will become a slave to the car. When you order parts, you will end up inspecting them for scratches and marks, and will probably end up buying something like a dozen mouldings just to get one that is not marked. Many dealers will not let you return specially-ordered parts. So when budgeting for building such a car, double your first estimate!

Restoring a 2002

Nuts and bolts of restoration

By this stage, you should know:

1. why you want to restore a 2002
2. which model you intend to restore
3. what parts you need
4. how much work you can undertake yourself
5. how much work you will need to farm out to professionals.

Try not to get too excited just yet; there is more planning to do before you can get started!

Plan your spending

Armed with your lists of parts to buy and work required, you can now plan how to spend your budget. It may well be that you do not have all the money you will need to complete the restoration when you start. If so, try to buy the parts you need as you progress through the work. For example, if you are going to use your car while you are carrying out some of the mechanical repairs and restoration, put money aside to buy the mechanical parts you need, and get the car mechanically sound before you start on the body and trim.

If possible, try to find out which parts may not be available in the future, and plan to buy these first. Then plan your purchases as you need them. That means buying the body panels and hard-to-find parts before buying the chrome trim and interior trim parts. Remember that mouldings and seat covers are difficult to store and are easily scratched or damaged, so buy these parts when you get towards the end of the project.

Dismantling

Most people seem to enjoy ripping cars apart, probably because for a little bit of work they seem to achieve a lot. But beware of going too fast. Make copious notes, even of blatantly obvious items, because you will certainly have forgotten how some thngs go back together by the time you start reassembling the car in six months' or a year's time. Good photographs are an essential, and even if you are not planning to sell the car they make a good record of the restoration and can show that the work was carried out properly.

Mechanical parts removal

Remove parts as a unit. A 2002 can be broken down into:

1 Bodyshell
2 Front subframe, complete with engine, gearbox, steering, front suspension and front brakes
3 Rear subframe, complete with final drive, rear suspension and rear brakes.

Always put nuts back onto bolts and make notes of which parts need repair or replacement. If you put all the nuts and bolts into one big box, you will end up having a lot left over at the end of your restoration! If nuts, bolts and clips go with an item, put them in a plastic bag and tape them to that item.

Restore in lumps

Tackle the repair of mechanical parts in sections. So, for example, you might deal with the front subframe, front suspension, steering and front brakes all together, and then move on to the rear subframe, rear suspension and rear brakes. At least then you will have complete restored units ready to fit to the bodyshell when that is completed.

Mechanical restoration

This is not the place to go into the finer points of mechanical restoration; workshop manuals already do this perfectly well. However, it is worth stressing here the value of that workshop manual. If you are carrying out a task on the car that you have never done before, read carefully everything the workshop manual says about it before you begin. If it says you need a special tool, obtain that tool before you start the job. Work cleanly, carefully and safely.

If you intend to rebuild your engine or to have a reconditioner rebore your cylinder block, make sure you have the new pistons before reboring because some oversizes are becoming difficult to find. Many engine reconditioners do not realise that pistons for BMW 2002 models are available in oversizes of +0.25mm as well as the normal oversizes of +0.50mm and +1.00mm. If your engine builder bores your block to accept +0.50mm pistons and the only pistons you can buy are +0.25mm, you have a problem!

Wiring loom

Use masking tape or sticky labels to identify wires and where they go to. If you think a sketch would help, draw one in your restoration notebook. With modern looms which have plastic insulated cables, tidying up is usually all that is needed. Replace damaged terminals with solder-on terminals if possible, and remember that you can always replace damaged areas of the outer loom covering with plastic loom tape (plain insulating tape without the sticky back).

Bodyshell restoration

There are three ways to tackle this. You can buy an exchange bodyshell, you can send the work out to a reputable bodyshop, or you can tackle the body repairs yourself. If you choose to do the job yourelf, you will need the space of at least two cars, plus suitable welding, grinding and metal cutting equipment.

The first stage is to strip out all the trim parts and sundry mechanical parts that are still left on the bodyshell. Label and make notes as you go. Next, you need to strip off the factory sound-proofing from the floor areas, both above and below the floor. Under the bitumen rust proofing and sound proofing you will usually find the restorer's nightmare - rust.

After that, you must decide which panels (or sections) you are going to replce. If sills or inner sills are among the items to be replaced, weld in temporary strengthening between A post, B post and transmission tunnel. You would be surprised how many doors do not fit because this was not done when the sills were replaced! When you are happy that the shell is not going to collapse or move, start removing the panels and sections that you are going to replace. Make sure you also remove old filler that may be hiding the evidence of "bodged" repairs.

A **2002** bodyshell stripped for full restoration. This picture makes clear where the complete front panel comes away from the inner wings.

Panel removal

Most panels fitted by the factory which are not bolted on are spot welded. The simplest and easiest way to remove these is to gently prise the two panels apart with a builder's bolster chisel. You will then be able to see where the spot welds are. Drill them out with a spot weld cutter, available from most motor factors.

Rust removal

This is a must. Rust is like a cancer, and it will spread underneath paintwork if it is not completely eradicated. The best method of rust removal is to have the bodyshell shot-blasted, but choose your shot-blaster carefully. Some individuals and companies are not used to working on the relatively delicate metal of a car body, and you may end up with distorted panels. This happens either because the panels have been blasted too hard, or because of heat build-up during the blasting process. If you are not blasting, cut out all rust and replace with new metal.

Don't bodge it!

This advice applies to most aspects of life, but to car restoration in particular. Carry out each step of the restoration or rebuild properly and you should not have to touch it again. Cut corners with structural corrosion and it will be back to haunt you like a ghost. Use only the best tools, parts and materials, and always use genuine or original equipment parts if they are still available.

Do not be a slave to originality for its own sake. Slight departures from standard may well keep the car on the road far longer and in much greater safety. In the 2002 floor pan, for example, you will find large plugs. These are always a rust trap and simply fill holes used to drain paint and chemicals when the bodies were dipped in the factory. You are better off without them.

A new spare wheel well panel has been welded in place here, but the wheelarch panels themselves were sound enough to be repaired.

Rust repairs

If you do not get rid of all the rust now, you will be restoring or repairing the bodywork again in six months or a year. This will be annoying, boring and expensive. Rust removal is the part of the restoration which can take the most time, and after days of grinding out rust it is easy to tell yourself that you have gone far enough and that a coat of rust-curing product should take care of the rest. Do not be tempted! Chemical engineering is good, and the magic products that turn rust into something else help, but they do not turn rust back into metal! The best method by far is to remove the rust completely.

Has the rust been removed?

If you are considering buying a restored BMW, first put yourself in the shoes of the person selling the car. Did he restore it to sell? If so, did he cut out all the rust, or did he cut corners? It may look bright and shiny now, but will it look this good in a year's time? If you bought a car to restore and sell, would you put huge amounts of time into removing rust and bad metal when you could easily hide it under filler and a few coats of paint? All these questions provide food for thought!

Shot blasting seems to be the best method of removing large areas of rust, but it must be carried out by somebody who is very skilled at his job. It is very easy to distort large panels by the build-up of too much heat. Blasted bodyshells need to be cleaned very thoroughly, and any grit must be blown or vacuumed out of all the crevices and box sections. The small bits of grit tend to hide themselves away until you come to put the final coat of paint on the car, and will then attach themselves to it!

Blasted shells or panels must be painted straight away with a good etching primer. Do not use any filler primers that are porous and let in moisture. Try to keep any primed parts dry and ventilated, or you will end up with micro-blisters to the paint finish later in the car's life.

Grinding is the best way of eradicating small areas of rust, and of course is also necessary for smoothing out welds. Use a good small electric or air-powered angle grinder. Again, when you have finished grinding or repairing the area you are working on, brush on your etching primer to protect the area so that rust will not start again.

Belt sanders are a must for places which are difficult to get at if you are not blasting. Their finger-like form makes it easy to get into areas where a grinder will not reach.

Disc sanders tend to score the metal and make filling and metal preparation difficult. The best way of dealing with light surface rust is to use an orbital sander with a maximum of 80-grit production paper.

Mule Skinners are essentially brushes which can be mounted on an arbour in the end of an electric or air drill. The wire brush is encapsulated in a resin. This means that the wires are always at the correct angle for removing the rust; ordinary rotary wire brushes tend only to remove rust for a short time until their wires bend. A Mule Skinner is a must for anyone who wants to repair or restore bodywork, as it removes rust without removing too much metal.

Rust-perforated metal can only be dealt with in one way. Do not be tempted to weld a plate over the affected area, but cut it out; plate-welding is MoT repair work, not restoration.

There are many methods of removing panel sections, and as many different tools to do the job. Among the most useful are:

1 Air saws: *These are tools with a saw blade at one end, that saw the metal by means of a reciprocating air motor which in turn makes the*

saw blade go backwards and forwards. Cheap ones are next to useless, but expensive ones work well and will last for many years.

2 Vibrating saws: *similar to the saws they use in hospitals to remove plaster casts. The saw has a round or semi-circular blade that vibrates. If you hold it against a panel, the blade will cut through the panel like a knife going through butter. Another wonderful thing about this type of saw is that when it comes into contact with a wiring loom (or human flesh), it does not cut! Its only drawback is that it will only cut straight lines; if you try to follow a curve, you will break the rather expensive blade.*

3 Hand operated snips: *These are excellent for getting blisters and for cutting small thin flat sheets of metal. Unfortunately, they are not much use in bodyshell restoration work.*

4 Air chisels: *Quick, but very noisy and not very accurate. If you do use an air chisel, make sure you wear strong gloves and goggles, and hope that you have deaf or very tolerant neighbours.*

Fit panels the proper way

Before welding new panels on to the bodyshell, fit all the panels to the shell temporarily by fixing either with small tack welds, or better still, with small self-tapping screws. As the original tooling for many of the BMW body panels is now old and worn, new panels may vary in size slightly from the original dimensions and will therefore not always fit perfectly. By checking the fit before welding you will be able to foresee any problems before it is too late.

When you are happy with the fits, remove each panel in turn. Strip or sand off the transit primer, apply weld through zinc primer to the areas that are to be spot welded, and spot weld and or MIG weld as required. When fitting part panels or repair sections it is best to "joggle" the edges of the joins leaving around a quarter to half inch overlap, and then to spot weld the panels together. Alternatively you can drill a hole through one sheet and "plug weld" with the MIG welder. Ideally the joined area should be lead-loaded before finishing with a thin skim of filler.

Many people tackling bodywork restoration wonder how the front wings are fitted to a 2002, because the method is not obvious. In fact, the

Restoring a 2002

When fitting new panels, it is important to check that they fit before beginning to weld them in place on the shell.

wings are bolted on, but the join to the front panel above the headlamp is leaded and finished smooth to leave no trace of a join.

Painting

No doubt it is possible to spray a car with a vacuum cleaner, but if you want the best results you had better not try it yourself! More seriously, if you do not have a suitable spray area and the proper equipment, leave paintwork to the professionals. You can have done a fantastic job with the bodywork, only to spoil the finished product by spraying with unsuitable material or in an unsuitable location. At the end of this book is Appendix D which gives information on the paint colours used on the 2002 models and the paint codes which paint suppliers will need if they are to supply paint for your car.

New wheelarch and sill panels in place on a **2002 TARGA** bodyshell under restoration.

Rustproofing

Rustproofing your restoration project is most important, especially if you live in a country where they "grit" the roads in winter with a mixture of salt and grit. Before rustproofing make sure that all the under sides of the body where panels join have been sealed properly with seam sealer and that adequate coats of anti-stone chip spray and paint have been applied.

First of all, spray cavity wax-based rustproofing product into inner sill areas, inside doors and any chassis rails or box sections. It is then an idea to spray the same material onto the underside panel joins before coating the complete underside in a modern wax-based rustproofing material.

Refitting parts

It is very easy to get carried away when you get to this stage in a restoration. The original good condition chrome that you considered suitable to re-use looks shabby against the newly painted body! As the parts are still available you decide to replace all the chrome parts including door handles and lamps, and so the story goes on. By the time you finish the restoration you will have spent over twice your budget. Your finished restoration will look immaculate, maybe even better than new, but will you want to use it? 2002 enthusiasts have varying ideas on how they wish to restore their car, and it is perfectly all right to prefer driving the car to worrying about how to keep it looking totally immaculate!

A useful tip when it comes to fitting the bright side mouldings is to drill any holes required before the shell is painted. Dip your plastic trim clips in cavity wax rustproofing material before fitting them to the car. After the car is complete, it is a good idea to go round it with an aerosol of rust proofer, spraying the material behind the mouldings, lamps and so on.

The restored car begins to look like a car once again …but there is still a long way to go!

Component detailing is important for an original appearance, and is much easier with the engine out.

The 2002 models have a particular problem with the fit of the door glass against the door shut rubbers. This is because the door drop glass has no frame at the top or rear. There are so many adjustments that can be made to alter the position of the glass that it is not unheard-of to spend eight hours to adjust one door and glass!

If the rubber seals for the front and screens are cracked and perished, change them now before the car starts to suffer from water leaks. It is difficult to get old, hard, windscreen rubbers to seal properly.

Upholstery and interior trim

As noted in an earlier chapter, the cloth material for the seats wears out quite quickly, although usually only the driver's seat is affected. Many of the original materials are still available from BMW and specialists. If you intend to recover just the driver's seat, do not expect the material to be a perfect match for that on the other seats. The original material will very probably have come from another batch, and the seats in your car will also have faded. If you want a perfect match, you will need to recover all the seats or perhaps cut some second hand material from an old back seat and use that to repair the worn driver's seat.

Carpets can be a problem with the Model 73 cars. None of the carpets are now available, and they were steam-moulded in one piece. The only way to get near to the original type is to get a coachtrimmer to sew the carpet into the car. If the trimmer is good, he can make the finished 'fitted' carpet look original.

Care for your restored 2002

You may be pleased to hear that on the whole the owner of a restored car looks after that car and lavishes it with more attention than if he or she had purchased a new car. If you have to store your car in a small garage, do make sure that the garage is well ventilated as in some instances it can be better for the car to be outside than be shut up in a humid garage.

Restoring a 2002

Wiring going into a freshly-restored **2002** bodyshell. Fitting new wiring without first stripping the interior is next to impossible!

BMW ALPINA
Castrol

Chapter Eleven

The 2002 in competition

By the time the 2002 came on-stream in 1968, BMW's sporting reputation was already a key feature of the company's image. Although the staid saloons and economy cars of the 1950s had swept away much of the sporting image which had been built up at the end of the 1930s with the legendary 328, the 1960s had seen a renaissance, first with the little rear-engined 700, then with the 1800TI and 1800TI/SA New Class four-door saloons, and latterly with the two-door 02 models. It was during 1967 that the 1600-2 had made its sporting debut, and it had shown such promise that the arrival of a 2-litre version of the same car was greeted with enormous enthusiasm by drivers and fans alike.

The 2002 proved an immediate success on the tracks in Europe, and over the next few years it also shone in rallying and in hillclimb events. The BMW works team was matched by teams from the aftermarket tuners Alpina, GS, Koepchen and Schnitzer, and there were also plenty of privateers who found the 2002 ideally suited to their racing budgets and aspirations. Between 1968 and 1972, there were few saloon car events of any consequence which did not feature 2002s heavily, and the car was widely recognised as the one to beat.

However, the 2002's dominance of European saloon car events could not and did not last. As quickly as the car had itself ousted the Alfa Romeos which had dominated events in the mid-1960s, so it was in turn ousted by the all-conquering Ford Escort after 1972. Although 2002s continued to compete and to show good returns right up to the end of their production run in 1976, they had by this time been outclassed.

Nevertheless, this was very far from being the end of the 2002's competitive career. With the rise of "historic" motorsport events in the 1980s, the 2002 rose to prominence once again. Today, it remains one of the most affordable cars for those whose interests lie in historic motorsport events, and it continues to give an excellent account of itself on the circuits.

The 2002 in competition.

1968

The 2002 entered European motorsport with a bang during 1968. The car proved sensationally successful, and by the end of the season BMW were able to boast that 2002s had won their categories in both the European Touring Car Championship and the European Hillclimb Championship. In addition, 2002s had claimed 101 outright victories in motorsport events, 152 category wins, no fewer than 1,099 class wins and 996 Gold Medals. That was certainly some going for a car's first season!

For 1968, the 2002s ran in Group 5 of the European Touring Car Championships. The factory's own drivers were Dieter Quester and Hubert Hahne, both of whom had driven the earlier works 1600-2 cars, and it was Quester who went on to become European Champion. Yet the 2002s did not demonstrate their prowess from Day One; in fact, the works cars were not ready in time to compete in the first round of the Championship at Monza, and injury prevented Hahne from driving in the second round in Vienna. Nevertheless, Quester pulled off a magnificent victory in Vienna - the car's first outing - and managed to set a new lap record for Touring Cars while he was at it. Third place at Snetterton, second in Brno, a win at the Nürburgring in Germany (when he switched to the second team car in mid-race) and a pair of third places at Zandvoort and Jarama were enough to give Quester the title.

Even at this stage, the BMW works team had a healthy relationship with the aftermarket tuning specialists. The second works car, driven by Dieter Basche in Vienna and by Hubert Hahne on other occasions, had been prepared by Schnitzer. It was also a Schnitzer-prepared 2002 which was successful in the European Hillclimb Championship, the driver in this instance being a BMW engineer called Ernst Furtmayr.

1969

The rules governing Group 5 in the European Touring Car Championship were fairly slack in the late 1960s: even the two-seater Porsche 911

The 2002 in competition.

somehow managed to qualify for the category, even though almost all the other cars were four-seater saloons! With the 911 in the running, BMW realised that they would have to develop something very special indeed if they were to keep the Championship for another year, and so they in turn exploited the rules by developing a turbocharged 2002.

The engine capacity of the 2002TIK remained standard so that the car could still compete in the 2-litre class, but the Eberspacher turbocharger boosted power from the 210bhp attainable with the standard engine to a thumping 280bhp. To maintain traction, the works team cars were fitted with fatter tyres (in fact Formula 2 types of 245 section at the front and 260 section at the rear), and to cover these they had purposeful-looking widened wheelarches.

The turbocharged cars did the trick. Starting with the short-distance events - their first long-distance event was the Brands Hatch Six-Hour Race in Britain - they racked up three convincing wins before the Porsche 911s were able to draw even by winning the next three events in the middle of the season. But Dieter Quester, who had claimed the first three victories, then won again at Jarama in September and thus clinched the Championship for the second year running.

Yet the works Porsches were not the only serious competition which the turbocharged 2002s had had to face. Very strong opposition had come from the Alpina team, racing highly-tuned 2002s which still had carburettors when both the factory and Schnitzer had turned to fuel injection; and so well had Alpina done that they managed to come second in the championship, ahead of the works Porsches. This was a convincing demonstration - if one were needed - of the effectiveness of the Alpina modifications.

Alpina 2002s also showed up well in the 2-litre class of the German Circuit Racing Championship during the 1969 season, driven by Jürgen Neuhaus, but it was Schnitzer who once again excelled. The Schnitzer-tuned 2002 driven by Ernst Furtmayr once again won its class in the European Hillclimb Championship, and Furtmayr also campaigned a Schnitzer-prepared 2002 to good effect in the German Circuit Racing Championship. The Koepchen team also raced Schnitzer-tuned 2002s, demonstrating their excellent reliability by winning 23 races out of the 26 for which they entered. And last but very far from least, the German Rally Championship for 1969 went to Helmut Bein, driving a 2002.

1970

There was a change in the regulations for the 1970 European Touring Car Championship. Turbocharged cars were no longer eligible for the Group 5 which the BMW works racers had won in 1969 unless at least 1,000 examples of the car were offered for sale. Failing that, they had to compete in the very much tougher Group 7. BMW were unable to meet the Group 5 requirements (although a 2002 turbo did go on sale some years later), and so the increased power of 290bhp from the 1970 turbocharged cars did them precious little good. As the 2002tii - the next most powerful 2002 model - was simply not competitive in Group 5, BMW fielded no works entries in that category for 1970.

AC Schnitzer were also a force to be reckoned with in **2002** racing. In this picture, a Schnitzer car leads the field at a corner.

Schnitzer, Koepchen and Alpina were still in there fighting, but they were unable to overcome the 2-litre Alfa Romeos and did not show well in the European Touring Car Championship. The German Touring Car Championship saw rather better results, when the young Hans Stuck (later Group C World Champion and a works driver for BMW, Ford and Porsche) won his class in the second round with a Koepchen 2002ti. However, it would have been difficult for the 2002 not to rack up some successes in the German Touring Car Championship for the very good reason that a very high proportion of the entrants were using 2002s! In the Nürburgring 24-hour race, for example, where Hans Stuck won his class, nine of the first ten cars to finish were 2002ti models!

Outside the circuits, the 2002 continued to prove a reliable rally entrant. Helmut Bein, this year partnered by Christoph Memel, won the German Rally Championship for the third year running with an Alpina-tuned 2002ti.

The 2002 in competition.

Above:
One of the very few Alpina Racers left, lovingly restored to its former glory.

Left:
Alpina were always one of the major names in preparing compition **2002s**. This example was based on a **2002ti**.

The 2002
in competition.

The 2002 in competition.

2002 Alpinas in action

The 2002 in competition.

1971

The year 1971 was not a good one for motor sport in general. A number of fatal accidents contributed to give it a negative image in Europe, and BMW works driver Hubert Hahne was among those who decided to stop racing after the death of his friend Jochen Rindt. The 2-litre class of the European Touring Car Championship was dominated by Alfa Romeos and by the new Ford Escorts, and the 2002s were not rewarded with victories despite some good showings by the Alpina, Schnitzer and Koepchen teams.

Nor did the 2002 perform well in the European Hill Climb Championship. This year, it was outclassed by the new Irmscher-tuned Opel Kadett.

The 2002 in competition.

1972

The Ford Escort 1600RS was the dominant force in the 2-litre class of the European Touring Car Championships for 1972. That year, the 2002s were outclassed, and in any case the thrust of the works racing effort had now passed to the big six-cylinder coupés.

However, the 2002 continued to prove a winner in rallies. Achim Warmbold, the 1971 World Rally Champion, drove his Alpina 2002 to third place in the Acropolis Rally of the World Rally Championship, and won the TAP Rally in Portugal, which counted towards the European Rally Championship. Closer to home, Rainer Zweibnumer won the German Rally Championship, also at the wheel of an Alpina 2002.

The 2002 in competition.

1973

The organisers of the European Touring Car Championship changed the rules once again for the 1973 season, this time mainly to prevent Ford from completely dominating the event with their Escorts and thus removing the interest from the championship. Four-valve engines were now permitted on condition that 100 examples of their special cylinder heads were built for sale. So, to keep pace with the four-valve works Escorts, BMW developed a four-valve version of their 2-litre engine.

Based on the earlier 1.6-litre Formula 2 racing engine, the BMW four-valve engine was good for 280bhp at 9,000rpm. Schnitzer also developed a four-valve racing conversion, which was slightly different from the BMW type. Throughout the 1973 season, Dieter Basche and Helmut Kelleners kept up the pressure on the works Fords, forcing them to fight hard for every victory; and the BMWs kept their end up so well that the championship was not decided in Ford's favour until late in the season. Dieter Basche had to be content with third place in the Championship, but the four-valve cars had demonstrated beyond doubt that the 2002 still had plenty to offer as a competition machine.

BMW also entered a pair of 2002ti models in the World Rally Championship for 1973, with drivers Achim Warmbold and Bjorn Waldegaard. Both cars used Schnitzer four-valve engines rather than BMW's own four-valve type because the Schnitzer engine gave more of the torque so necessary in rallying.

1974

The end of 1973 was blighted by the oil crisis, and motorsport inevitably came under very public scrutiny. For 1974, BMW judged it prudent to present the image of a responsible manufacturer, and declined to field a works team. Ford also pulled out of competition after the first round of the European Touring Car Championship. However, both BMWs and Fords remained on the circuits to relive some of the excitement of the previous year's battles.

The competition cars did not represent the works teams, of course. Ford's colours were flown by the Zakspeed team, which raced Escorts; and the leading BMW exponents this year were GS. Their team leader was Dieter Basche, but it was his team-mate Jorg Obermoser who took the GS 2002 to second place overall in the European Touring Car Championship. The winner, though, was a Zakspeed Escort.

The 2002 in competition.

This GS-tuned **2002** won the German Touring Car Championship in the mid-1970s in the hands of Jorg Denzel. It is now in private ownership and was pictured at a classic car show.

The 2002 in competition.

1975

The 2002s were seen on the circuits again, mainly in the German Touring Car Championship. However, the best they could achieve was fourth in class, in the hands of Jörg Obermoser; Fords won again. In the German Hill Climb Championship, Walter Struckmann achieved fourth place overall in his 2002.

1976

The 2002 range ceased production in 1976, and the 1976 season was to be the last one in which they would participate with factory blessing. The car was by this stage no longer a front runner, although it did achieve some good placings. Jörg Obermoser started the season well with a win at the Nürburgring in the first round of the European Touring Car Championship, but could achieve nothing better than seconds and thirds as the season wore on.

Meanwhile, in the German Circuit Racing Championship, Andreas Schall came second overall with a Schnitzer 2002. Walter Struckmann once again came fourth overall with his 2002 in the German Rally Championship.

The 2002 in competition.

1977

Paradoxically, 1977 proved to be another interesting year for the 2002 in competition. The new 320 had taken over as the leading light of the BMW works team, but the regulations for the European Touring Car Championship had changed once again to allow turbocharged cars, and Schnitzer developed a 2002 to suit. This was a remarkable machine, which mustered 400bhp from a turbocharged downsized 1.4-litre edition of the 2-litre BMW engine. Unfortunately, the car did not live up to its promise, and was not seen again for the following season.

Above: Schnitzer's 1977 competitions entry was a 400bhp turbocharged **2002** - but with its engine reduced in capacity to 1.4 litres.

Right: Now in private hands, the ex-Schnitzer **2002 TURBO** shows off its widened rear track.

The 2002 in competition.

1978-1979

As the 1970s drew to a close, privateers continued to enter 2002s in motor sport events, though without spectacular success. In racing terms, the cars were now old, and it is a measure of their abilities that so many remained in use for so long. However, the professionals turned to newer models, and the numbers of 2002s in mainstream competition work dwindled to nothing over the next few years.

Eyecatching colour schemes remain the order of the day. This is racing **2002** prepared by Jaymic for BMW Car Club events and driven by the company's Keith Morley.

The BMW **2002** continues to figure in classic saloon racing. This German example is typical of the breed.

The 2002 in competition.

Multiple colours and multiple carburettors: the engine bay of a modern **2002** racer.

401,947

Appendices
a. Vehicle Identification
b. Paint Colours
c. Technical Specifications

Appendices

Appendix a: Vehicle Identification (chassis or commission number)

On a 2002 these will be found on a plate towards the rear of the right hand front inner wing. The chassis number is also stamped into the right hand inner wing panel at the rear edge. On USA specification cars, the VIN number is also repeated on the top surface of the instrument panel just inside the windscreen. The engine number would originally have been the same as the chassis number and is stamped on the engine crank case just above the starter motor. The following list will help to identify any 2002 and to give an approximate year of its build date.

	Model	Chassis No. Range	Dates of Production	Quantity	Total
2002 RHD	Pre 71	1 650 001 to 1 653 979	Jan '68 to March '71	3,979	
	Model 71	1 653 980 to 1 655 000	April '71 to April '73	1,021	
	Model 71	2 550 001 to 2 551 473	April '73 to Aug '73	1,473	
	Model 73	4 200 001 to 4 203 228	Aug '73 to Oct '75	3,228	9,701
2002 LHD	Pre 71	1 600 001 to 1 650 000	Jan '68 to March '70	50,000	
	Pre 71	2 600 001 to 2 634 875	March '70 to March '71	34,875	
	Model 71	2 634 876 to 2 650 000	April '71 to Oct '71	15,125	
	Model 71	3 600 001 to 3 652 610	Oct '71 to Aug '73	52,610	
	Model 73	3 660 001 to 3 700 000	Aug '73 to March '75	40,000	
	Model 73	3 750 001 to 3 754 244	March '75 to June '75	4,244	196,854
2002 RHD (USA)	Pre 71	1 660 001 to 1 680 000	Feb '68 to Jan '71	20,000	
	Pre 71	2 570 001 to 2 572 069	Jan '71 to March '71	2,069	
	Model 71	2 572 070 to 2 593 704	April '71 to Aug '73	21,635	
	Model 73	4 220 001 to 4 230 601	Sept '73 to July '74	10,601	
	Model 75	2 360 001 to 2 367 750	July '74 to July '75	7,750	
	Model 76	2 370 001 to 2 380 000	Sept '75 to Feb '76	10,000	
	Model 76	2 740 001 to 2 745 584	Feb '76 to July '76	5,584	77,639
2002 AUTO RHD	Pre 71	2 520 001 to 2 520 993	Jan '69 to Jan '72	993	
	Model 71	2 520 994 to 2 522 160	Jan '72 to Aug '73	1,167	
	Model 73	4 270 001 to 4 271 240	Aug '73 to Oct '75	1,240	3,400
2002 AUTO LHD	Pre 71	2 500 001 to 2 507 130	Jan '69 to March '71	7,130	
	Model 71	2 507 131 to 2 513 880	April '71 to Aug '73	6,750	
	Model 73	4 250 001 to 4 255 109	Aug '73 to June '75	5,109	18,989
2002 AUTO LHD (USA)	Pre 71	2 530 001 to 2 532 124	May '69 to June '70	2,124	
	Model 71	2 532 126 to 2 534 861	Jan '72 to Aug '73	2,736	
	Model 73	4 280 001 to 4 283 429	Sept '73 to July '74	3,429	
	Model 75	2 380 001 to 2 382 692	Aug '74 to June '75	2,692	
	Model 76	2 390 001 to 2 393 292	Sept '75 to July '76	3,292	14,272
2002ti LHD	Pre 71	1 680 001 to 1 696 448	Sept '68 to April '71	16,448	16,448
2002tii RHD	Model 71	2 750 001 to 2 752 447	June '71 to Aug '73	2,447	
	Model 73	2 770 001 to 2 771 233	Aug '73 to Oct '75	1,233	3,680

Appendices

	Model	Chassis No. Range	Dates of Production	Quantity	Total
2002tii LHD	Model 71	2 700 001 to 2 720 114	April '71 to Aug '73	20,114	
	Model 73	2 730 001 to 2 737 460	Aug '73 to June '75	7,460	27,574
2002tii LHD (USA)	Model 71	2 760 001 to 2 764 521	Aug '71 to Aug '73	4,521	
	Model 73	2 780 001 to 2 782 928	Sept '73 to Dec '74	2,928	7,449
2002 CABRIOLET LHD	Pre 71	2 790 001 to 2 790 200	Jan '71 to June '71	200	200
2002 TARGA RHD	Model 71	2 791 001 to 2 791 260	Jan '73 to July '73	260	
	Model 73	3 595 001 to 3 595 094	Sept '73 to Jan '74	94	354
2002 TARGA LHD	Model 71	2 795 001 to 2 796 200	July '71 to July '73	1,200	
	Model 73	3 590 001 to 3 590 763	Aug '73 to Dec '75	763	1,963
2000/02 TOURING* RHD	Model 71	3 440 001 to 3 441 537	Sept '71 to Aug 73	1,537	
	Model 73	3 480 001 to 3 480 339	Aug '73 to April '74	339	1,876
	*N.B. RHD TOURING MODELS FROM 3 441 034 WERE BADGED 2002 FROM DEC 72				
2000/02 TOURING* LHD	Model 71	3 350 001 to 3 360 264	April '71 to Aug '73	10,264	
	Model 73	3 460 001 to 3 462 840	Aug '73 to April '74	2,840	13,104
	*LHD TOURING MODELS FROM 3 357 720 WERE BADGED 2002 FROM DEC 72				
2000 AUTO TOURING* LHD	Model 71	3 400 001 to 3 400 989	May '71 to Sept '72	989	989
2000/02tii TOURING* LHD	Model 71	3 420 001 to 3 425 059	April '71 to Aug '73	5,059	
	Model 73	3 430 001 to 3 430 724	Aug '73 to April '74	724	5,783
	*N.B. NO FIGURES AVAILABLE FOR THE BADGE CHANGE TO 2002tii. CHANGE FROM DEC 72				
2002 TURBO LHD	Model 73	4 290 001 to 4 291 672	Jan '74 to June '75	1,672	1,672

2002:	284,194	**2002 TURBO:**	1,672
2002 TOURING:	14,980	**2002 TARGA:**	2,317
2002 TOURING AUTO:	989	**2002 CABRIOLETS:**	200
2002 TII TOURING:	5,783	**2002ti:**	16,448
2002 AUTO:	36,6612	**2002tii:**	38,703

Total:401,947

Appendices

Appendix b: Paint Colours

The paint colours used on the 2002 models had German names and were identified by a paint code number. The following alphabetical list contains all the German names, together with their code numbers and a colour reference.

Colour	Years	Code
Agave DARK GREEN	70 - 74	071
Amazonagrün MID GREEN	74 - 75	076
Anthrazitgrau MET DARK GREY	74 - 79	055
Arktisblau MET DARK BLUE	74 - 86	045
Atlantik BLUE	68 - 74	041
Baikal MET MID BLUE	71 - 74	042
Bristol BLUE	65 - 66	058
Ceylon MET GOLD	72 - 74	008
Chamonix WHITE	62 - 80	085
Colorado LIGHT ORANGE	68 - 73	002
Derby	62 - 69	706
Fjord MET LIGHT BLUE	72 - 80	037
Florida LIGHT GREEN	59 - 71	066
Golf YELLOW	71 - 79	070
Granada MID RED	63 - 72	023
Granatrot MET LIGHT RED	74 - 76	025
Inka DARK ORANGE	71 - 75	022
Jadegrun MID GREEN	75 - 78	078
Korall RED	76 - 80	020
Madeira MAROON	76 - 78	027
Malaga MAROON	70 - 76	021

Colour	Years	Code
Manila IVORY	62 - 79	104
Mintgrün PASTEL GREEN	75 - 78	079
Nevada BROWN-GREY	68 - 72	001
Pacific DARK BLUE	64 - 68	500
Pastelblau MID BLUE	75 - 78	04
Phonix DARK ORANGE	76 - 77	028
Polaris SILVER METALLIC	63 - 76	060
Resedagrun MET LIGHT GREEN	76 - 81	075
Riviera DARK BLUE	70 - 74	036
Rubinrot MET LIGHT RED	76 - 80	018
Sahara BEIGE	67 - 75	006
Schwarz BLACK	70>	086
Sienabrun LIGHT BROWN MET	74 - 77	009
Sierra Beige YELLOW BEIGE	75 - 78	012
Taiga MET GREEN	72 - 76	072
Tampico LIGHT BEIGE	64 - 69	707
Topasbraun MET DARK BROWN	76 - 80	013
Tundra MET MID GREEN	71 - 73	068
Türkis MET TURQUOISE	72 - 75	065
Verona BRIGHT RED	72 - 75	024

Appendices

SALOONS & CABRIOLETS
1968-1971

Solid	Metallic
Agave	Baikal
Atlantic	Ceylon
Bristol	Fjord
Chamonix	Polaris
Colorado	Taiga
Florida	Tundra
Granada	
Inca	
Nevada	
Riviera	
Sahara	
Schwarz[70>]	

SALOONS & TOURING
1971-1972

Solid	Metallic
Agave	Baikal
Atlantic	Ceylon
Bristol	Fjord
Chamonix	Granatrot
Colorado	Polaris
Florida	Taiga
Sienabraun	Tundra
Golf	
Granada	
Inca	
Nevada	
Riviera	
Sahara	
Schwarz[70>]	
Verona	

SALOONS & TOURING
1972-1973

Solid	Metallic
Agave	Baikal
Amazonasgrün	Ceylon
Atlantic	Fjord
Bristol	Granatrot
Colorado	Polaris
Florida	Sienabraun
Golf	Türkis
Granada	Taiga
Inca	Tundra
Nevada	
Riviera	
Sahara	
Schwarz[70>]	
Verona	

SALOONS
1974-1976

Solid	Metallic
Agave	Anthrazitgrau
Amazonasgrün	Baikal
Atlantic	Ceylon
Bristol	Fjord
Colorado	Granatrot
Florida	Polaris
Golf	Sienabraun
Granada	Türkis
Inca	Taiga
Mintgrün[75>]	
Nevada	
Pastellblau[74>]	
Riviera	
Sahara	
Schwarz[70>]	
Verona	

TOURING
1974-1976

Solid	Metallic
Agave	Baikal
Amazonasgrün	Ceylon
Atlantic	Fjord
Bristol	Granatrot
Colorado	Polaris
Florida	Sienabraun
Golf	Türkis
Granada	Taiga
Inca	
Nevada	
Riviera	
Sahara	
Schwarz[70>]	
Verona	

TURBO
1974-1975

Solid	Metallic
Chamonix	Polaris

TARGA
1971-1974

Solid	Metallic
Amazonasgrün[72>]	Anthrazitgrau[74>]
Chamonix	Arktisblau[76]
Granada	Granatrot
Jadegrün[75>]	Polaris
Mintgrün[75>]	
Pastellblau[74>]	

Appendices

Appendix c: Technical Specifications

2002
2002 AUTOMATIC
2000/2002 TOURING
2002 CABRIOLET
2002 TARGA

Engine	Four-cylinder 1,990cc with five-bearing crankshaft and chain-driven overhead camshaft. Iron block and alloy head. 89mm bore x 80mm stroke; compression ratio 8.5:1 (European models), 8.3:1 (US models for 1972 and 1975-1976); single Solex 40PDSI carburettor. Maximum power 100PS at 5,500rpm (European models), 113bhp SAE at 5,800rpm (US models to 1974), 98bhp at 5,500rpm (US models for 1975-1976). Maximum torque 116 lb/ft at 3,000rpm (European models and US models to 1974), 106 lb/ft at 3,500rpm (US models for 1975 -1976).

Transmission	2002 Four-speed - All-synchromesh gearbox with single dry plate clutch.							
	Gear ratios:	(1) 3.764:1	(2) 2.020:1	(3) 1.320:1	(4) 1.000:1		(R) 4.096:1	Axle ratio: 3.64:1
	Optional five-speed - All-synchromesh gearbox with single dry plate clutch.							
	Gear ratios:	(1) 3.368:1	(2) 2.160:1	(3) 1.579:1	(4) 1.241:1	(5) 1.000:1	(R) 4.000:1	Axle ratio: 3.64:1
	2002 Auto - Three-speed ZF automatic gearbox with torque converter.							
	Gear ratios:	(1) 2.56:1	(2) 1.52:1	(3) 1.00:1			(R) 2.00:1	Axle ratio: 3.64:1

Suspension, steering and brakes	Independent front suspension with coil springs, MacPherson struts and anti-roll bar. Independent rear suspension with coil springs, semi-trailing arms and anti-roll bar. (Front and rear anti-roll bars optional on US-model 2002 Automatic). Worm and roller steering with 15.5:1 ratio. Front disc brakes and rear drums, with servo assistance.

Dimensions	98.4" (2,500mm)		
Wheelbase	'68-'71: 52.4" (1,330mm)	'71-'75: 53" (1,348mm)	US saloons: '74-'76: 52.8" (1,340mm)
Front track	'68-'71: 52.4" (1,330mm)	'71-'75: 53" (1,348mm)	US saloons: '74-'76: 52.8" (1,340mm)
Rear track	Saloon & cabriolet: 170" (4,230mm)	Touring: 162" (4,115mm)	US saloons: '74-'76: 176" (4,470mm)
Overall length	62.6" (1,590mm)		
Overall width	Saloon: 55.5" (1,410mm)	Cabriolet: 53.5" (1,360mm)	Touring: 54" (1,372mm)
Overall height	saloon: 2183lb (990kg)	Auto saloon: 2227lb (1010kg)	Targa: 2293 lb (1040kg)
Unladen weight	cabriolet: 2337lb (1060kg)	Touring: 2337lb (1060kg)	US model saloon: 2400 lb (1088kg)

Appendices

2002TI

Engine	Four-cylinder 1,990cc with five-bearing crankshaft and chain-driven overhead camshaft. Iron block and alloy head. 89mm bore x 80mm stroke; compression ratio 9.3:1; two Solex 40PHH carburettors. Maximum power 120PS at 5,500rpm. Maximum torque 123 lb/ft at 3,600rpm.
Transmission	Four-speed all-synchromesh gearbox with single dry plate clutch. Gear ratios (1) 3.835:1 (2) 2.053:1 (3) 1.345:1 (4) 1.000:1 (R) 4.180:1 Axle ratio 3.65:1
	Optional five-speed all-synchromesh gearbox with single dry plate clutch. Gear ratios (1) 3.368:1 (2) 2.160:1 (3) 1.579:1 (4) 1.241:1 (5) 1.000:1 (R) 4.000:1 Axle ratio 3.65:1
Suspension, steering and brakes	Independent front suspension with coil springs, MacPherson struts and anti-roll bar. Independent rear suspension with coil springs, semi-trailing arms and anti-roll bar. Worm and roller steering with 15.5:1 ratio. Front disc brakes and rear drums, with servo assistance.

Dimensions	
Wheelbase	98.4" (2,500mm)
Front track	53" (1,348mm)
Rear track	53" (1,348mm)
Overall length	170" (4,230mm)
Overall width	62.5" (1,590mm)
Overall height	56" (1,410mm)
Unladen weight	2072lb (940kg)

Appendices

2002 TURBO

Engine	Four-cylinder 1,990cc with five-bearing crankshaft and chain-driven overhead camshaft. Iron block and alloy head. 89mm bore x 80mm stroke; compression ratio 6.9:1; Schäfer PL-04 mechanical fuel injection with KKK type BLD turbocharger operating at 7psi. Maximum power 170PS at 5,800rpm; maximum torque 179 lb/ft at 4,000rpm.
Transmission	Four-speed all-synchromesh gearbox with single dry plate clutch. Gear ratios (1) 3.351:1 (2) 1.861:1 (3) 1.279:1 (4) 1.000:1 (R) 3.650:1 Axle ratio 3.36:1
	Optional five-speed all-synchromesh gearbox with single dry plate clutch. Gear ratios (1) 3.368:1 (2) 2.160:1 (3) 1.579:1 (4) 1.241:1 (5) 1.000:1 (R) 3.650:1 Axle ratio 3.36:1
Suspension, steering and brakes	Independent front suspension with coil springs, MacPherson struts and anti-roll bar. Independent rear suspension with coil springs, semi-trailing arms and anti-roll bar. Worm and roller steering with 12.8:1 ratio. Front ventilated disc brakes and rear drums, with servo assistance.

Dimensions	
Wheelbase	98.4" (2,500mm)
Front track	53.6" (1,362mm)
Rear track	53.6" (1,362mm)
Overall length	166.1" (4,220mm)
Overall width	63.7" (1,620mm)
Overall height	56i" (1,410mm)
Unladen weight	2282lb (1035kg)

Further reading for BMW 2002 Owners

Restore it right the first time!

This information-packed guide provides all the information you need to restore your prized BMW.

Covers the 1600 to 2002 turbo models produced from the sixties through to the mid seventies.

Detailed information on restoring bodywork, engine and trim, etc., plus production changes, paint schemes and history by Mike Macartney.

With a total of 224 fully illustrated pages. Soft Bound.

ISBN: 9781855204515

The '02 began as the brainchild of Max Hoffmann, BMWs US importer, who wanted a car to follow up the 1600-2. So popular did it become that it went on to sell more than 300,000 examples over the seven seasons of its production.

The story of the 2002 is told through 64 contemporary articles covering road and comparison tests, specifications, technical data and drivers impressions.

Models covered: 2002, Ti, Tii, Turbo and Alpina.

With a total of 208 fully illustrated pages. Soft Bound.

ISBN: 9781855207370

All of Brooklands Books titles are available from Amazon and good motoring books shops.

www.brooklandsbooks.com

The Authors

James Taylor is one of the leading British writers on motoring history, with more than two dozen books to his credit. He writes for several magazines in Britain and in continental Europe, and has also written for the US press. He is in demand by motoring manufacturers as an analyst and adviser on motoring matters.

Mike Macartney is one of the owners of Jaymic, the independent BMW specialists based in Cromer, Norfolk, England. He has an unrivalled knowledge of the 02-series cars and he and his company are respected worldwide for their skills with these classis BMWs. Through Jaymic, Mike has also been closely involved in competition with 2002s, and has developed some highly regarded conversions for these cars.